W9-CDV-079

PAINLESS
Geometry

Lynette Long, Ph.D.

illustrated by Tom Kerr

BARRON'S

JUN

All inquiries should be addressed to:
Barron's Educational Series, Inc.
250 Wireless Boulevard
Hauppauge, New York 11788
http://www.barronseduc.com

Library of Congress Catalog Card No.: 2001029516

ISBN-13: 978-0-7641-1773-2
ISBN-10: 0-7641-1773-4

Library of Congress Cataloging-in-Publication Data

Long, Lynette.
    Painless geometry / Lynette Long ; illustrated by Tom Kerr.
        p.   cm.
    Includes index.
    ISBN 0-7641-1773-4
    1. Geometry—Juvenile literature. [1. Geometry.]   I. Kerr, Tom, ill. II. Title.

    QA445.5 .L66   2001
    516—dc21

                                                            2001029516

PRINTED IN CANADA
20 19 18 17 16 15 14 13 12

CONTENTS

CONTENTS

INTRODUCTION

This book is designed to make geometry painless. It has several unique features to help you both enjoy geometry and excel at it.

Experiments

The book contains a series of experiments designed to help deepen your understanding of geometry. If you do the experiments, you will gain an intuitive understanding of geometry. Why not try some of the experiments? They are fun and informative.

Painless Steps

Complex procedures are divided into a series of painless steps. These painless steps help you solve problems in a systematic way. Just follow the steps one at a time, and you'll be able to solve most geometry problems.

Examples

Most problems are illustrated with an example. Study each of the examples. If you are in trouble, copy the example and learn it. Research shows that writing/copying the problem and the solution may help you understand it.

Illustrations

Painless Geometry is full of illustrations to help you better understand geometry. Geometry is a visual subject. It's important for you to draw the problems and examples, too. If you can visualize a problem, it will be easier to understand it.

Caution—Major Mistake Territory

Occasionally you will find boxes entitled "Caution—Major Mistake Territory." These boxes will help you avoid common pitfalls. Be sure to read them carefully.

Study Strategies

Occasionally you will see sections
called "Study Strategies." Here
you will find tips on how to study
the material in a particular section
that are guaranteed to improve your performance.

Mini-Proofs

Numerous mini-proofs appear throughout the
book. Written informally, these mini-proofs
are the first step toward understanding formal
geometry proofs.

Brain Ticklers

There are three or four problems after each section of the book.
These exercises are designed to make sure that you understand
what you just learned. Complete all the Brain Ticklers. If you get
any wrong, go back and study the materials and examples before it.

Super Brain Ticklers

At the end of each chapter are Super Brain Ticklers, which will
review all the material in the chapter. Solve these problems to
make sure that you understand each chapter.

Don't move forward to the next section of the book until you
really understand what was in the last chapter. Geometry is a linear subject, and what you don't know in one chapter will haunt
you in the next. Don't move forward if any topic confuses you.
Go back and try to figure out what you don't know.

Geometry can be painless, so why not get started?

Chapter One is titled "A Painless Beginning," and it really is. It
is an introduction to the basic terms and symbols of geometry.

Chapter Two, "Angles," will teach you everything about angles. It
will show you how to measure and classify different types of angles
as well as explain the relationship between different angle pairs.

Chapter Three is titled "Parallel and Perpendicular Lines" and
will show you how to identify parallel and perpendicular lines.
You will learn what types of angles are formed when parallel and
perpendicular lines are formed.

Chapter Four shows you that there is more to learn about triangles than you ever dreamed possible. You will learn the types of triangles, as well as how to find the perimeter and area of a triangle.

Chapter Five explains what similar and congruent triangles are. By the time you finish reading this chapter, you will know what SSS, SAS, ASA, and AAS mean and how to use each of them to prove that two triangles are congruent.

Chapter Six is titled "Quadrilaterals," which is a fancy word for four-sided figures. Trapezoids, parallelograms, rhombuses, rectangles, and squares are all quadrilaterals. You will learn all about these interesting polygons here.

Chapter Seven could have been titled "Circles, Circles, and More Circles." You may know what a circle is, but you will find out how a pie is not only something you eat but an essential term in geometry.

Chapter Eight will show you how to find the perimeter, area, and volume of some common and not so common shapes. It's painless, just watch.

Chapter Nine, "Graphing," will teach you how to graph a point and a line. You can use what you learn here to draw your own geometric shapes.

If you are learning geometry for the first time, or if you are trying to remember what you learned but forgot, this book is for you. It is a painless introduction to geometry that is both fun and instructive. Dive in—and remember, it's painless!

Web Addresses Change!

You should be aware that addresses on the World Wide Web are constantly changing. Although the addresses provided were current when this book was written, sooner or later, some of the addresses may no longer work. If you should come across a web address (URL) that no longer appears to be valid, either because the site no longer exists or because the address has changed, either shorten the URL to the first slash or do a key word search using geometry and related terms. For example, type in geometry and then narrow the search by typing in pi or even "homework help." Don't forget the quotes!

WARNING: Not every response to your search will match your criteria, and some sites may contain adult material. If you are ever in doubt, check with someone who can help you.

CHAPTER ONE

A Painless Beginning

Geometry is a mathematical subject, but it is just like a foreign language. In geometry you have to learn a whole new way of looking at the world. There are hundreds of terms in geometry that you've probably never heard before. Geometry also uses many common terms, but they have different meanings. A point is no longer the point of a pencil, and a plane is not something that flies in the sky. To master geometry you have to master these familiar and not so familiar terms, as well as the theorems and postulates that are the building blocks of geometry. You have a big job in front of you, but if you follow the step-by-step approach in this book and do all the experiments, it can be painless.

UNDEFINED TERMS

Undefined terms are terms that are so basic they cannot be defined, but they can be described.

A *point* is an undefined term. A point is a specific place in space. A point has no dimension. It has no length, width, or depth. A point is represented by placing a dot on a piece of paper. A point is labeled by a single capital letter.

A

A *line* is a set of continuous points that extend indefinitely in either direction. In this book, the term *line* will always mean a straight line.

This line is called line *AB*, or line *AC*, or line *BC*, or $\overleftrightarrow{AB}$, $\overleftrightarrow{AC}$, or $\overleftrightarrow{BC}$. The two-ended arrow over the letters tells you that the letters are describing a line. You can name a line by any two points that lie on the line.

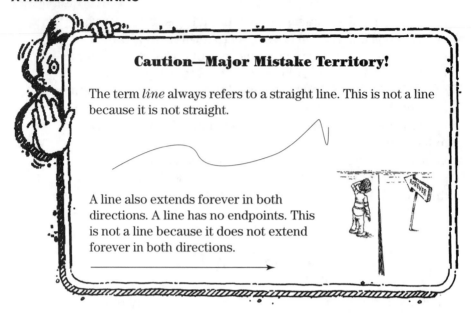

Caution—Major Mistake Territory!

The term *line* always refers to a straight line. This is not a line because it is not straight.

A line also extends forever in both directions. A line has no endpoints. This is not a line because it does not extend forever in both directions.

A *plane* is a third undefined term. A plane is a flat surface that extends infinitely in all directions. If you were to imagine a table-top that went in every direction forever, you would have a plane.

A plane is represented by a four-sided figure. Place a capital letter in one of the corners of the plane to name the plane. This is plane *P*.

SOME DEFINED TERMS

A line segment is part of a line with two endpoints. A line segment is labeled by its endpoints. We use two letters with a bar over them to indicate a line segment. The line segment is segment XY or $\overline{XY}$.

A *ray* has one endpoint and extends infinitely in the other direction. The endpoint and one other point on the ray label a ray. This ray could be labeled ray *AB* or ray *AC*. When you label a ray, put a small arrow on the top of the two letters to indicate that it is a ray. The arrow on top of the two letters should show the direction of the ray. This ray could also be labeled or $\overrightarrow{AB}$ or $\overrightarrow{AC}$.

Opposite rays are two rays that have the same endpoint and form a straight line. Ray *BA* and ray *BC* are opposite rays. We can also say that $\overrightarrow{BA}$ and $\overrightarrow{BC}$ are opposite rays.

Parallel lines are two lines in the same plane that do not intersect.

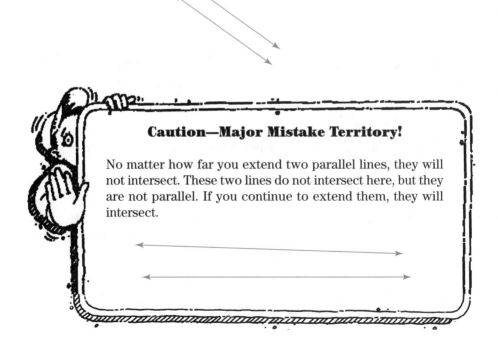

Caution—Major Mistake Territory!

No matter how far you extend two parallel lines, they will not intersect. These two lines do not intersect here, but they are not parallel. If you continue to extend them, they will intersect.

Collinear points lie on the same line. A, B, and C are collinear points.

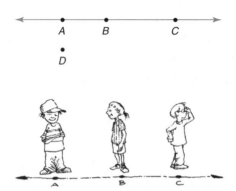

Noncollinear points do not lie on the same line. X, Y, and Z are noncollinear points. They cannot lie on the same line.

Y

Z

X

BRAIN TICKLERS
Set # 1

Decide whether each of the following statements is true or false.

1. A ray has one endpoint.

2. Points that lie on the same line are called linear.

3. A line is always straight.

4. A line segment can be ten miles long.

(Answers are on page 19.)

MORE DEFINED TERMS

An *angle* is a pair of two rays that have the same endpoint.

The rays are the *sides* of the angles. The endpoint where the two rays meet is called the *vertex* of the angle.

A *polygon* is a closed figure with three or more sides that intersect only at their endpoints. The sides of the polygons are line segments. The points where the sides of a polygon intersect are called the vertices of the polygons.

These are all polygons.

Two shapes are *congruent* if they have exactly the same size and shape.

Each of these pairs of shapes is congruent.

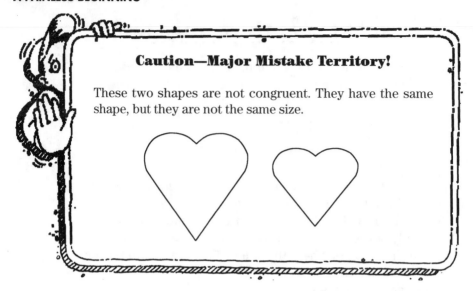

Caution—Major Mistake Territory!

These two shapes are not congruent. They have the same shape, but they are not the same size.

The *perimeter of a polygon* is the sum of the measures of all the sides of the polygon. The perimeter of a polygon is expressed in linear units such as inches, feet, meters, miles, and centimeters.

The *area* of a geometric figure is the number square units the figure contains. The area of a figure is written in square units such as square inches, square feet, square miles, square centimeters, square meters, and square kilometers.

The *ratio* of two numbers a and b is a divided by b written $\dfrac{a}{b}$ as long as b is not equal to zero.

A *proportion* is an equation that sets two ratios equal. $\dfrac{a}{b} = \dfrac{c}{d}$ is an example of a proportion.

BRAIN TICKLERS
Set # 2

Determine whether each of the following statements is true or false.

1. A polygon can be composed of three segments.

2. The perimeter of a figure can be measured in square inches.

3. 5/0 is a ratio.

(Answers are on page 19.)

POSTULATES

Postulates are generalizations in geometry that cannot be proven true. They are just accepted as true. Do the following experiment to discover one of the most basic postulates of geometry.

Experiment

Find out how two points determine a line.

Materials
Pencil
Paper
Ruler

Procedure

1. Draw two points on a piece of paper. To draw a point make a dot on the paper as if you were dotting the letter i.
2. Draw a line through these two points. Now draw another line through these same two points. Draw a third line through these two points. Are the three lines the same or different?
3. Draw another two points. How many different lines can you draw through these two points?

Something to think about . . .

How many lines can you draw through a single point?

Postulate: Two points determine a single line.

What does this postulate mean?

- You can draw many lines through a single point.
- You can only draw one line through any two different points.

Experiment

Find out how many points determine a plane.
Materials
Clay or play dough
Four sharpened pencils with erasers
Tabletop

Procedure

1. Form a ball of clay the size of a small orange.
2. Stick the pointed end of two sharpened pencils one inch into the ball of clay.

3. Turn the ball of clay so that both erasers touch the table at the same time.
4. Remove the pencils and place them point first into the ball of clay in a different way. Turn the ball so that both erasers touch the table at the same time.
5. Stick a third pencil into the ball of clay.
6. Turn the ball of clay so that all three erasers touch the table at the same time.
7. Remove all three pencils from the ball of clay and place them point first in the ball of clay in a different way. Find a way to touch all three erasers to the table at the same time.
8. Now stick four sharpened pencils in the ball of clay. Is it possible to turn the ball of clay so that all four erasers touch the table at the same time?
9. Remove all four pencils and place them point first in the ball of clay in a different way. Is there a way to touch all four erasers to the table at the same time?
10. Place the four pencils in the ball of clay so that it is impossible for all four erasers to touch the table at the same time.

Something to think about . . .

How many legs must a stool have so that it does not tip over?
How many legs must a stool have to guarantee it always makes solid contact with the floor?
Did you ever have a wobbly chair? Why was that?

Postulate: At least three points not on the same line are needed to determine a single plane.

What does this postulate mean?

- Through a single point there are an infinite number of planes.
- Through two points you can also find an infinite number of planes.

- Through three points that lie in a straight line there are an infinite number of planes.
- Through three points that do not lie in a straight line there is only one plane.

POSTULATE: IF YOU ARE LATE YOU WILL MISS THE PLANE.

Postulate: If two planes intersect, they intersect on exactly one line.

What does this postulate mean?

- Not all planes intersect.
- Two different planes cannot intersect on more than one line.

THEOREMS

Theorems are generalizations in geometry that can be proven true.

Theorem: If two lines intersect, they intersect at exactly one point.

What does this theorem mean?

- Not all lines intersect.
- Two different lines cannot intersect at more than one point.

This statement is a theorem and not a postulate because mathematicians can prove this statement. They do not have to accept it as true.

> **Theorem:** Through a line and a point not on that line, there is only one plane.

Three noncollinear points determine a single plane. A line provides two points and the point not on that line is the third noncollinear point.

GEOMETRIC PROOFS

There are two types of proofs, *inductive* proofs and *deductive* proofs.

Inductive proofs

Inductive proofs start with examples, and then these examples are used to support a conclusion.

5 is an odd number.
15 is an odd number.
25 is an odd number.
35 is an odd number.
Conclusion: If a number ends in a five, it is an odd number.

Deductive proofs

Deductive proofs are the classic geometric proofs. Deductive reasoning uses definitions, theorems, and postulates to prove a new theorem true. A typical deductive proof uses a two-column format. The statements are on the left and the reasons are on the right.

STATEMENT	REASON

In this book, deductive proofs will have a different format. Three questions will be used to construct the proof. These three questions will help you think about proofs in a painless way.

1. What do you know?

2. What can you infer based on what you know?

3. What can you conclude?

GEOMETRIC SYMBOLS

Much of geometry is written in symbols. You have to understand these symbols if you are going to understand geometry. Here are some common geometric symbols.

$=$ equal to
$3 + 5 = 8$
Three plus five is *equal to* eight.

$\neq$ not equal to
$6 \neq 2 + 1$
Six is *not equal to* two plus one.

>   greater than
> $7 > 5$
> Seven is *greater than* five.

$\geq$ greater than or equal to
$4 \geq 4$
Four is *greater than or equal to* four.

<   less than
$6 < 10$
Six is *less than* ten.

$\leq$ less than or equal to
$1 \leq 1$
One is *less than or equal to* one.

$\triangle$ triangle
$\triangle ABC$
Triangle *ABC*

$\cong$ congruent to
$\triangle ABC \cong \triangle DEF$
Triangle ABC is *congruent to* triangle *DEF*.

$\perp$ perpendicular to
line $AB \perp$ line CD.
$\overleftrightarrow{AB} \perp \overleftrightarrow{CD}$
Line *AB* is *perpendicular to* line *CD*.

segment $XY \perp$ segment YZ
$\overline{XY} \perp \overline{YZ}$
Segment *XY* is *perpendicular to* segment *YZ*.

ray $AB \perp$ ray CD
$\overrightarrow{AB} \perp \overrightarrow{CD}$
Ray *AB* is *perpendicular to* ray *CD*.

π   pi
area of a circle = πd

The area of a circle is *pi* times the diameter.

~   similar to
△ *ABC* ~ △ *DEF*
Triangle *ABC* is *similar to* triangle *DEF*.

↔   line
$\overleftrightarrow{AB}$
Line *AB*

→   ray
$\overrightarrow{AB}$
Ray *AB*

—   segment
$\overline{AB}$
Segment *AB*

⌒   arc
$\overparen{AB}$
Arc *AB*

∥   parallel lines
WX ∥ *YZ*
Line *WX* is parallel to line *YZ*.

∠   angle
∠ *A*
Angle *A*

BRAIN TICKLERS
Set # 3

Name the following symbols.

1. ≅
2. <
3. ⊥
4. ≥

(Answers are on page 19.)

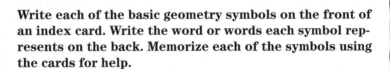

Study Strategies

Write each of the basic geometry symbols on the front of an index card. Write the word or words each symbol represents on the back. Memorize each of the symbols using the cards for help.

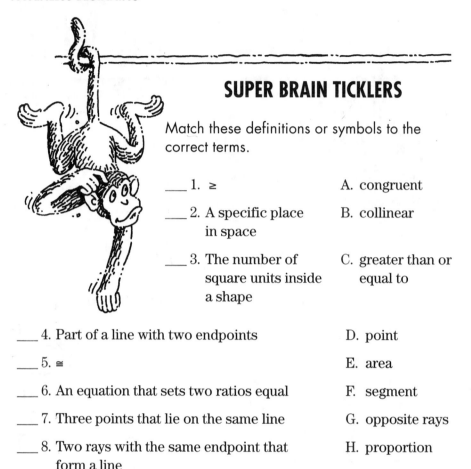

SUPER BRAIN TICKLERS

Match these definitions or symbols to the correct terms.

___ 1. ≥   A. congruent

___ 2. A specific place   B. collinear
       in space

___ 3. The number of   C. greater than or
       square units inside   equal to
       a shape

___ 4. Part of a line with two endpoints   D. point

___ 5. ≅   E. area

___ 6. An equation that sets two ratios equal   F. segment

___ 7. Three points that lie on the same line   G. opposite rays

___ 8. Two rays with the same endpoint that   H. proportion
       form a line

                                                  I. equal to

(Answers are on page 19.)

BRAIN TICKLERS—THE ANSWERS

Set # 1, page 6

1. True
2. False
3. True
4. True

Set # 2, page 9

1. True
2. False
3. False

Set # 3, page 17

1. Congruent to
2. Less than
3. Perpendicular to
4. Greater than or equal to

Super Brain Ticklers, page 18

1. C
2. D
3. E
4. F
5. A
6. H
7. B
8. G

CHAPTER TWO

Angles

An angle is two *rays* with the same endpoint that do not lie on the same line. The common endpoint is the vertex of the angle. The rays are the sides of the angles. These two rays have the same endpoint.

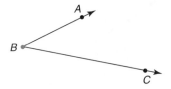

The endpoint *B* is the vertex of the angle. The sides of the angle are rays *BA* and *BC*.

Label an angle using three letters, one letter from one side, one letter from the other side, and the letter of the vertex. Remember that the letter of the vertex is in the middle. Label the above angle ∠*ABC* or ∠*CBA*.

Caution—Major Mistake Territory!

Be sure to put the letter of the vertex in the *middle* of the three letters.

This is angle *XYZ*, or you could label it angle *ZYX*, but it is not angle *XZY* or angle *YXZ*. The *Y* must go in the middle of the three letters.

An angle can also be labeled using the letter of the vertex. Angle *XYZ* could just be labeled angle *Y*. Be careful, you can only use one letter to label an angle if no other angle shares the same vertex. This is angle *M*.

Perhaps the easiest way to label an angle is to put a number inside its vertex. When you use a number label, make a small arc inside the angle to make sure there is no confusion.

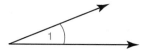

You can label a single angle four different ways.

Angle *WXY*
Angle *YXW*
Angle *X*
Angle 2

Caution—Major Mistake Territory!

If two angles share the same vertex, you cannot label them using a single letter.

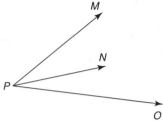

Which angle is angle *P*?
There are three angle *P*s.
　　　Angle *MPN*
　　　Angle *NPO*
　　　Angle *MPO*
To differentiate between the three angles, you must use three letters to label each of them or you can use numbers.

An angle divides a plane into three distinct regions:

- interior of the angle
- angle itself
- exterior of the angle

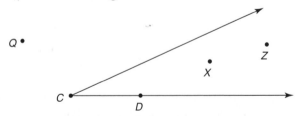

Points *X* and *Z* lie in the interior of the angle
Points *C* and *D* lie on the angle
Point *Q* lies in the exterior of the angle

BRAIN TICKLERS
Set # 4

Determine whether each of the following statements is true or false.

1. Angles can be labeled in more than one way.
2. Angle *ABD* is the same as angle *DBA*.

3. The measure of angle 2 as marked is equal to the measure of angle *ABD*.

(Answers are on page 46.)

MEASURING ANGLES

You measure the length of a line segment using a ruler. You measure the size of an angle using a *protractor*. The unit of measure for an angle is the *degree*. A protractor is a semicircle that is marked off in 180 increments. Each increment is one degree.

When expressing the measure of an angle, write "the measure of angle *ABC* is 25 degrees." This is abbreviated as m∠*ABC* = 25. The symbol for degrees is °, but it is generally omitted when measurements are written in the abbreviated format. However, if the measurement is given in the angle itself, the degree sign is used.

Caution—Major Mistake Territory!

Don't write ∠*B* = 25, since ∠*B* is not 25 degrees. The measure of angle *B* is 25 degrees.

Don't write m∠*ABC* = 25°. The degree sign is assumed!

Using a protractor

Notice that there are four things about the protractor here.

1. There are numbers from 0 to 180.

2. The numbers are marked from right to left and from left to right around the curved part of the semicircle.

3. If the protractor is made of clear plastic, there is a horizontal line along the base of the protractor. The clear plastic line intersects the protractor at 0 and 180 degrees.

4. There is a mark at the halfway point of the straight side of a protractor.

To use a protractor, follow these painless steps.

Step 1: Place the center marker over the vertex of the angle.

Step 2: Place the horizontal line along the base of the protractor on one of the sides of the angle.

Step 3: Read the number where the other side of the angle intersects the protractor. This is the measure of the angle.

Watch as these angles are measured.

What is the measure of this angle?

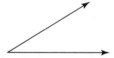

It measures 30 degrees.

Place the protractor on this angle.

Notice that the angle measures 45 degrees.

Measure this angle. It measures 120 degrees.

This angle is greater than 180 degrees. The small arc indicates that the angle is the large angle, which goes all the way around the small angle.

To measure an angle greater than 180 degrees, follow these painless steps.

Step 1: Measure the small angle.

Step 2: Subtract the measure of the small angle from 360.

EXAMPLE:
Measure this angle.

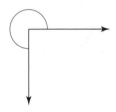

Step 1: Measure the small angle.
The measure of the small angle is 90 degrees.

Step 2: Subtract the measure of the small angle from 360.

$$360 - 90 = 270$$

The measure of the large angle is 270 degrees.

Caution—Major Mistake Territory!

When you measure angles, you are measuring the number of degrees in the angle, not the length of the sides. The length of the sides does not matter. Look at these two angles. Which is larger? Angle *A* or angle *B*?

Angle *B* is larger than angle *A*.
Angle *A* has longer sides, but angle *B* contains more degrees.

BRAIN TICKLERS
Set # 5

Measure the following angles.

1. 15

2. 150

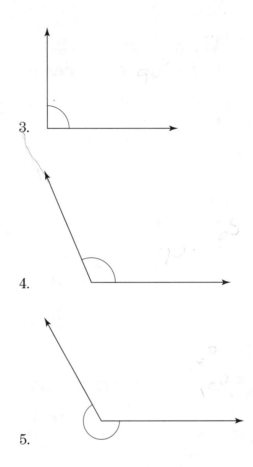

3.

4.

5.

(Answers are on page 46.)

CLASSIFYING ANGLES

Once you know how to measure angles you can classify them.

An *acute angle* measures less than 90 degrees. These are all acute angles.

A *right angle* measures exactly 90 degrees. These are both right angles.

An *obtuse angle* measures more than 90 degrees but less than 180 degrees. These are all obtuse angles.

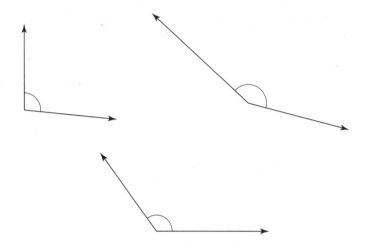

A *straight angle* measures exactly 180 degrees. This is a straight angle.

A *reflex angle* has a measure greater than 180 degrees and less than 360 degrees. These are reflex angles.

BRAIN TICKLERS
Set # 6

Decide whether each of these statements is true or false.

1. A 63-degree angle is an acute angle. *true*

2. A 90-degree angle is a straight angle. *false*

3. A 179-degree angle is an obtuse angle. *true*

4. A 240-degree angle is an obtuse angle. *false*

5. A 130-degree angle is a reflex angle. *false*

(Answers are on page 46.)

ANGLE PAIRS

If two angles have a total measure of 90 degrees they are called *complementary angles*.

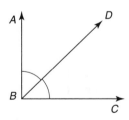

Angles *ABD* and *DBC* are complementary.
If the measure of angle *ABD* is 40 degrees, then the measure of angle *DBC* is 50 degrees.

$$40 + 50 = 90$$

To find the complement of an angle, use these two painless steps.

Step 1: Find the measure of the first angle.

Step 2: Subtract the measure of the first angle from 90 degrees.

EXAMPLE:
Find the complement of a 10-degree angle.

Step 1: Find the measure of the first angle. It is given as 10 degrees.

Step 2: Subtract the measure of the first angle from 90 degrees.

$$90 - 10 = 80$$

The complement of a 10-degree angle is an 80-degree angle.

Theorem: If two angles are complementary to the same angle, the measures of the angles are equal to each other.

EXAMPLE:
If $m\angle A = 25$
and $\angle B$ is complementary to $\angle A$
and $\angle C$ is complementary to $\angle A$,
then $m\angle B = m\angle C$.
Because $\angle B$ is complementary to $\angle A$, $m\angle B$ is $(90 - 25)$ or 65.
Because $\angle C$ is complementary to $\angle A$, $m\angle C$ is $(90 - 25)$ or 65.

If two angles have a total measure of 180 degrees they are called *supplementary angles*.

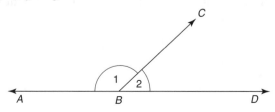

EXAMPLE:

> Angles *ABC* and *CBD* are supplementary.
> If angle 1 measures 125 degrees, angle 2 must measure 55
> degrees, since 125 + 55 = 180.

To find the supplement of an angle, use these two painless steps.

Step 1: Find the measure of the first angle.

Step 2: Subtract the measure of the first angle from 180
degrees.

EXAMPLE:

Find the supplement of a 10-degree angle.

Step 1: Find the measure of the first angle. It is given as 10
degrees.

Step 2: Subtract the measure of the first angle from 180
degrees.

$$180 - 10 = 170$$

The supplement of a 10-degree angle is a 170-degree angle.

> **Theorem:** If two angles form a straight line, the angles
> are supplementary.

EXAMPLE:

Angle *ABD* and angle *DBC* form a straight angle. Angle *ABD* and
angle *DBC* are supplementary to each other.

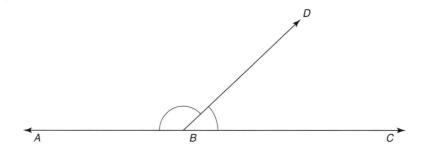

Theorem: If two angles are supplementary to the same angle, the measures of the angles are equal to each other.

EXAMPLE:

If m∠A = 70
and ∠B is supplementary to ∠A
and ∠C is supplementary to ∠A,
then m∠B = m∠C.
Because ∠B is supplementary to ∠A, m∠B is (180 – 70) or 110.
Because ∠C is supplementary to ∠A, m∠C is (180 – 70) or 110.

An *angle bisector* is a ray that divides an angle into two equal angles.

BISECTOR

EXAMPLE:

Ray *BD* bisects angle *ABC*. If angle *ABC* measures 60 degrees, then angle *ABD* measures 30 degrees, and angle *DBC* also measures 30 degrees.

30 + 30 = 60

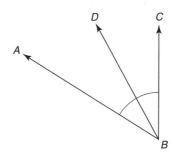

Four angles are formed when two lines intersect. The opposite angles always have the same measure, and they are called *vertical angles*.

EXAMPLES:
Angles 1 and 3 are vertical angles. Angles 2 and 4 are also vertical angles. The measure of angle 1 is equal to the measure of angle 3. The measure of angle 2 is equal to the measure of angle 4.

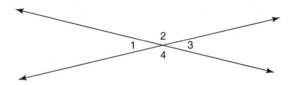

If angle 1 measures 70 degrees, then angle 3 must also measure 70 degrees.
If angle 2 measures 110 degrees, then angle 4 must also measure 110 degrees.

Two angles are *adjacent* if they share a common side and a common vertex but do not share any interior points.

EXAMPLE:

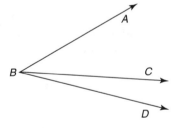

Angle *ABC* is adjacent to angle *CBD*. They share side *BC*.
Angle *ABD* is *not* adjacent to angle *CBD*. They share side *BD*, but they have common interior points.

Caution—Major Mistake Territory!

Angle *ABD* is *not* adjacent to angle *ABC*. They share side *AB*, but they have common interior points.
Angle *ABD* is *not* adjacent to angle *CBD*. They share side *BD*, but they have common interior points.

BRAIN TICKLERS
Set # 7

1. Angle *A* measures 59 degrees, what is the measure of its complement?

2. If angle *B* measures 12 degrees, what is the measure of its supplement?

3. If angle *C* measures 50 degrees and it is bisected, what is the measure of the two resulting angles?

Look at these angles.

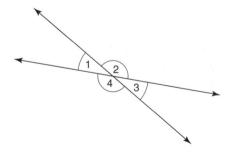

4. If angle 1 measures 30 degrees, what is the measure of angle 2?

5. If angle 1 measures 30 degrees, what is the measure of angle 3?

6. If a straight angle is bisected, what is the measure of each of the resulting angles?

(Answers are on page 46.)

ANGLE CONGRUENCE

Two figures with exactly the same size and shape are congruent. How do you determine if two angles are congruent? Two angles are congruent if any of the following conditions are met:

HEY! WE'RE THE SAME!

- They have the same measure.
- You can put one on top of the other, and they are identical.
- They are both right angles. Two right angles both measure 90 degrees. Since two right angles have the same measure, they are congruent.

- They are complements of the same angle. Two complements of the same angle always have the same measure.
- They are supplements of the same angle. Supplements of the same angle always have the same measure.
- They are vertical angles. Vertical angles are always congruent.
- They are both congruent to the same angle. If two angles are congruent to the same angle, they have the same measure as that angle and they must be equal to each other.

Caution—Major Mistake Territory!

Make sure that you know the difference between *equal* and *congruent*.
If m∠1 = 32 and m∠2 = 32, angle 1 is not equal to angle 2. Two angles cannot be equal. Angles are congruent but not equal.
Angle 1 is congruent to angle 2. However, the measure of angle 1 can be equal to the measure of angle 2.

BRAIN TICKLERS
Set # 8

Look at the following diagram. Which angles are congruent to each other?

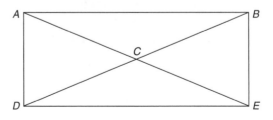

(Answers are on page 47.)

WORD PROBLEMS

It's possible to use your skills in algebra and what you are learning about geometry to solve word problems.

EXAMPLE:

If an angle is twice as large as its complement, what is the measure of the angle and its complement?

Change this problem to an equation.

The phrases in parentheses are changed to mathematical language.

x	(an angle)
=	(is)
2	(twice)
$90 - x$	(its complement)

Now write an equation.

$x = 2(90 - x)$

Solve this equation.

$x = 2(90 - x)$

$x = 180 - 2x$

Add $2x$ to both sides.

$3x = 180$

$x = 60$

The angle is 60 degrees; its complement is $90 - 60$ or 30 degrees.

EXAMPLE:

An angle is 50 degrees less than its supplement. What is the measure of the angle and its supplement?

Change the problem to an equation. Watch as the phrases in parentheses are changed to mathematical language.

x (an angle)

$=$ (is)

$180 - x$ (it's supplement)

-50 (50 degrees less than)

Now put all these expressions together to form an equation.

$x = 180 - x - 50$

Now solve this equation.

First simplify both sides of the equation.

$x = 180 - x - 50$

Simplified this equation becomes $x = 130 - x$.
Simplify both sides of the equation further to get $2x = 130$.
Divide both sides by 2, and the result is $x = 65$.
The angle is 65 degrees and its supplement is $180 - x$ or 115 degrees.

SUPER BRAIN TICKLERS

How well do you know your geometry terms?
Fill in the blanks with angle language.

1. _right_____ angles measure exactly 90 degrees.

2. The sum of two ___90°___ angles is exactly 180 degrees.

3. Two angles with exactly the same measure are _____.

4. An angle with a measure of less than 90 degrees is a(n) _____ angle.

5. The sum of two _____ angles is exactly 90 degrees.

6. _____ angles are formed when two lines intersect.

7. _____ angles measure greater than 90 degrees and less than 180 degrees.

8. An angle with a measure greater than 180 degrees and less than 360 degrees is called a(n) _____ angle.

9. A ray that divides an angle into two congruent angles is called an angle _____.

10. An angle that measures exactly 180 degrees is a(n) _straight_ angle.

(Answers are on page 47.)

BRAIN TICKLERS—THE ANSWERS

Set # 4, page 26

1. True

2. True

3. True

Set # 5, page 31

1. 15°

2. 150°

3. 90°

4. 110°

5. 240°

Set # 6, page 35

1. True

2. False

3. True

4. False

5. False

Set # 7, page 40

1. 31°

2. 168°

3. 25°, 25°

4. 150°

5. 30°

6. 90°, 90°

Set # 8, page 42

Angle *ACD* ≅ angle *BCE*
Angle *ACB* ≅ angle *DCE*
Angle *BED* ≅ angle *EDA* ≅ angle *DAB* ≅ angle *ABE*

Super Brain Ticklers, page 45

1. Right

2. supplementary

3. congruent

4. acute

5. complementary

6. Vertical

7. Obtuse

8. reflex

9. bisector

10. straight

CHAPTER THREE

Parallel and Perpendicular Lines

Points, lines, and planes are the primary elements of geometry. This chapter will explore the relationship between the lines in a plane. If two lines are in a plane, either they intersect or they are parallel.

Experiment

Discover the properties of lines in a plane.

Materials
    Two pencils
    A table

Procedure

1. Place two pencils on a table. The table represents a plane.
2. Let the two pencils represent two straight lines, but remember that straight lines extend infinitely. Make an X with the pencils. The pencils now intersect at exactly one point.

3. Try to make the pencils intersect at more than one point. It's impossible! The only way to make two pencils intersect at more than one point is to put them on top of each other, but then they represent the same line.

Something to think about . . .
    How many different ways can two lines intersect?

> **Theorem:** If two lines intersect, they intersect at exactly one point.

EXAMPLE:

These lines intersect at exactly one point, *A*.

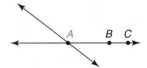

However, two lines can intersect an infinite number of different ways. These lines could also intersect at point *B*, point *C*, or any other point along either line.

If two lines intersect they form four angles.

When two lines intersect, the sum of any two adjacent angles is 180 degrees.

EXAMPLE:

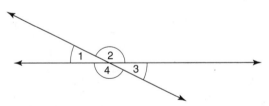

Angle 1 and angle 2 are adjacent angles, so their sum is 180 degrees.
Angle 1 and angle 4 are adjacent angles, so their sum is 180 degrees.
Angle 2 and angle 3 are adjacent angles, so their sum is 180 degrees.
Angle 3 and angle 4 are adjacent angles, so their sum is 180 degrees.

The sum of all four angles formed by two intersecting lines is 360 degrees.

EXAMPLE:

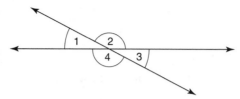

The sum of angles 1, 2, 3, and 4 is 360 degrees.

VERTICAL ANGLES

If two lines intersect, they form four angles. The opposite angles formed are called *vertical angles.*

EXAMPLE:

Angles *a* and *c* are vertical angles. Angles *b* and *d* are vertical angles.

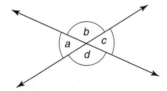

Experiment

Explore the relationship between vertical angles.

Materials
    Pencils
    Protractor

Procedure

1. Cross two pencils.
2. Look at the vertical angles. Use the protractor to measure them. Do the angles have the same measure? Are they congruent?

3. Move the pencils to create four new angles. Measure them. Are the vertical angles congruent?

Something to think about . . .
Is there a way to intersect two lines so that the vertical angles are not congruent?

Theorem: Vertical angles are always congruent.

Theorem: Vertical angles always have the same measure.

EXAMPLE:

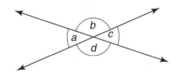

Angles *a* and *c* are vertical angles; therefore . . .
  angles *a* and *c* are congruent.
  angles *a* and *c* have the same measure.
Angles *b* and *d* are vertical angles; therefore . . .
  angles *b* and *d* are congruent.
  angles *b* and *d* have the same measure.

BRAIN TICKLER
Set # 9

Use this diagram to solve the problems that follow.

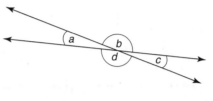

1. If m∠a = 15, what is the measure of angle b?
2. If m∠a = 15, what is the measure of angle c?
3. If m∠a = 15, what is the measure of angle d?
4. What is the sum of the measures of angles a and b?
5. What is the sum of the measures of angles b and c?
6. What is the sum of the measures of angles c and d?
7. What is the sum of the measures of angles d and a?
8. What is the sum of the measures of angles a, b, c, and d?

(Answers are on page 71.)

PERPENDICULAR LINES

Sometimes two lines intersect to form right angles.

Theorem: Perpendicular lines always intersect to form four right angles.

EXAMPLE:

These lines are perpendicular to each other.

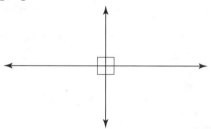

All four of these angles are right angles. The sum of all four of these angles is 90 + 90 + 90 + 90 = 360. The opposite angles are vertical angles and are congruent. All four angles are congruent to each other.

Experiment

Discover the number of lines that can be drawn from a point not on the line.

Materials
    Red pencil
    Black pencil
    Ruler
    Paper

Procedure

1. Draw a red line on a piece of paper.
2. Use the black pencil to place a black point above the red line.
3. How many lines can you draw through the black point that also intersect the red line?

Something to think about . . .
    How many lines can be drawn perpendicular to the red line that go through a single point?

Theorem: Given a line and a point not on that line, there is exactly one line that passes through the point perpendicular to the line.

Experiment

Explore the relationship between a line and a point on the line.

Materials
　　Red pencil
　　Black pencil
　　Paper
　　Ruler

Procedure
　　1. Draw a red line on a piece of paper. Make a black dot on the red line.

　　2. How many lines can you draw through the black dot?
　　3. Draw a second red line on a piece of paper. Make a black dot on this line.
　　4. How many perpendicular lines can you draw through the black dot?

Something to think about . . .
　　How many perpendicular lines can you draw through two parallel lines?

Theorem: Through a point on a line, there is exactly one line perpendicular to the given line.

Theorem: If two lines intersect and the adjacent angles are congruent, then the lines are perpendicular.

Mini-Proof

How can you prove the theorem: *If two lines intersect and the adjacent angles are congruent, then the lines are perpendicular? What do you know?*

1. Angle 1 and angle 2 are adjacent angles.

2. Angle 1 and angle 2 are congruent.

3. The measure of angle 1 plus the measure of angle 2 is equal to a straight angle, which is 180.

What can you infer based on what you know?

4. Since angles 1 and 2 are congruent, the measures of angles 1 and 2 are equal.

5. Since angle 1 is equal to angle 2 and they form a straight angle, they must each be equal to 90 degrees.

6. Since angle 1 and angle 2 are each 90 degrees, they are each right angles.

What can you conclude?

7. Since both angle 1 and angle 2 are right angles, the lines must be perpendicular.

BRAIN TICKLER
Set # 10

Determine whether each of these statements is true or false.

1. Two intersecting lines are always perpendicular. *false*

2. Two different lines can intersect at more than one point. *true* *false*

3. Perpendicular lines form acute angles. *false*

4. The sum of the measures of the angles of two perpendicular lines is 360 degrees. *true* *true*

5. All right angles are congruent to each other. *true*

(Answers are on page 71.)

PARALLEL LINES

Two lines are parallel if and only if they are in the same plane and they never intersect. Parallel lines are the same distance from each other over their entire lengths.

Experiment

Find parallel lines.

Materials
    Five pencils
    Paper

Procedure

1. Lay a pencil on a piece of paper. This piece of paper represents a plane.
2. Place a second pencil on the piece of paper parallel to the first pencil.
3. Place a third pencil on a piece of paper parallel to the other two pencils.
4. Place a fourth pencil on the paper parallel to the other three pencils.
5. Place a fifth pencil on the paper parallel to the other four pencils.

Something to think about . . .
    How many lines are parallel to any line?

Theorem: Given a line and a point not on the line, there is only one line parallel to the given line.

Draw a red line. Draw a black dot, not on the line. Label it *P*.

• *P*

How many lines can you draw through the black dot parallel to the red line?

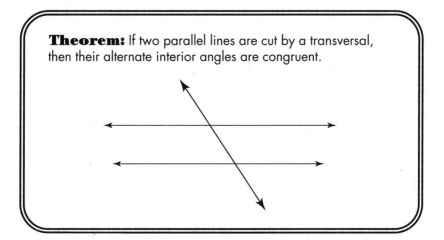

Theorem: If two parallel lines are cut by a transversal, then their alternate interior angles are congruent.

What does the term *transversal* mean?

> A *transversal* is a line that intersects two lines at different points.

What does the term *alternate interior* mean?

> *Alternate* means on different sides of the transversal.
> *Interior* means between the two parallel lines.
> *Alternate interior* means between the two parallel lines and on opposite sides of the transversal.

EXAMPLE:

Look at these two parallel lines.

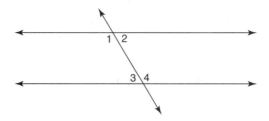

Angles 1 and 4 are alternate interior angles.
Angles 2 and 3 are alternate interior angles.
The measure of angle 1 is equal to the measure of angle 4.
The measure of angle 2 is equal to the measure of angle 3.

When two parallel lines are cut by a transversal, eight differ-ent angles are formed. If you know the measure of one of the eight angles, you can find the measure of all eight of the angles.

EXAMPLE:

If the measure of angle 3 is 60 degrees, you can find the measure of all the rest of the angles.

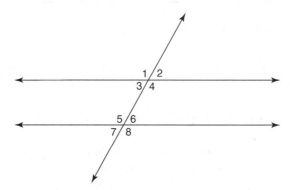

The measure of angle 2 is also 60 degrees, since it is vertical to angle 3.
The measure of angle 6 is also 60 degrees, since it is an alter-nate interior angle to angle 3.
The measure of angle 7 is also 60 degrees since it is a vertical angle to angle 6.

The measure of angle 1 is 120 degrees since it is supplemental to angle 3.

The measure of angle 4 is 120 degrees, since it is vertical to angle 1.

The measure of angle 5 is 120 degrees, since it is an alternate interior angle to angle 4.

The measure of angle 8 is 120 degrees, since it is a vertical angle to angle 5.

BRAIN TICKLERS
Set # 11

Assume that the measure of angle 4 is 110 degrees.

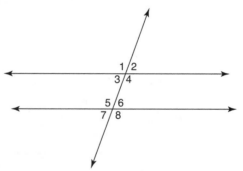

1. What is the measure of angle 1?

2. What is the measure of angle 2?

3. What is the measure of angle 3?

4. What is the measure of angle 5?

5. What is the measure of angle 6?

6. What is the measure of angle 7?

7. What is the measure of angle 8?

(Answers are on page 71.)

MORE ON PARALLEL LINES

> **Theorem:** If two parallel lines are cut by a transversal, then their corresponding angles are congruent.

What are *corresponding angles*? Corresponding angles are angles that lie on the same side of the transversal. One corresponding angle lies on the interior of the parallel lines while the other corresponding angle lies on the exterior of the parallel lines.

EXAMPLE:

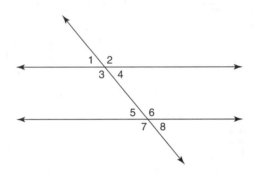

Angles 1 and 5 are corresponding angles, so they are congruent.
Angles 2 and 6 are corresponding angles, so they are congruent.
Angles 3 and 7 are corresponding angles, so they are congruent.
Angles 4 and 8 are corresponding angles, so they are congruent.

> **Theorem:** If two parallel lines are cut by a transversal, alternate exterior angles are congruent.

Exterior angles are found above or below the pair of parallel lines.
Alternate angles lie on opposite sides of the transversal.
Alternate exterior angles lie outside the parallel lines, and on the opposite sides of the transversal.

EXAMPLE:

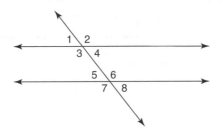

Angles 1 and 8 are alternate exterior angles; therefore, they are congruent.
Angles 2 and 7 are alternate exterior angles; therefore, they are congruent.

Theorem: If two parallel lines are cut by a transversal, the consecutive interior angles are supplementary.

The *consecutive interior angles* are angles of the same side of the transversal and inside both parallel lines.

EXAMPLE:

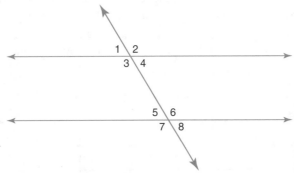

Angles 3 and 5 are consecutive interior angles.
Angles 4 and 6 are consecutive interior angles.
Angles 3 and 5 are supplementary angles.
Angles 4 and 6 are supplementary angles.

Experiment

Explore the relationship between angles formed by parallel lines and a transversal.

Materials
  Pencil
  Paper
  Ruler
  Protractor

Procedure

1. Draw two parallel lines.
2. Draw a transversal across the lines.
3. Label the eight angles formed angles 1, 2, 3, 4, 5, 6, 7, and 8.
4. Measure each of the angles and write its measure on the diagram.
5. Describe the relationship between each pair of angles as complementary, supplementary, equal, and unknown. Enter the results in the chart. Place a *C* for Complementary, *S* for Supplementary, *E* for Equal, and *U* for Unknown.

	Angle 1	Angle 2	Angle 3	Angle 4	Angle 5	Angle 6	Angle 7	Angle 8
Angle 1								
Angle 2								
Angle 3								
Angle 4								
Angle 5								
Angle 6								
Angle 7								
Angle 8								

Something to think about . . .
  What did you notice about the relationship of the angles?

Sum it up!

When two parallel lines are cut by a transversal, the following pairs of angles are congruent:

- alternate interior angles
- corresponding angles
- alternate exterior angles

When two parallel lines are cut by a transversal, the following pairs of angles are supplementary:

- consecutive interior angles

> **Theorem:** If two parallel lines are cut by a transversal, then any two angles are either congruent or supplementary.

Any two angles formed by two parallel lines are never complementary.

> **Theorem:** If a line is perpendicular to one of two parallel lines, then it is perpendicular to the other.

Proving lines parallel

There are three ways to prove two lines parallel. First, cut the two lines by a transversal. If any of the following are true, then the lines are parallel.

1. The alternate interior angles are congruent.

2. Their corresponding angles are congruent.

3. Their alternate exterior angles are congruent.

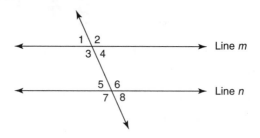

Way 1

If angle 4 and angle 5 are congruent, then line *m* is parallel to line *n*.

If angle 3 and angle 6 are congruent, then line *m* is parallel to line *n*.

Way 2

If angle 1 and angle 5 are congruent, then line *m* is parallel to line *n*.

If angle 2 and angle 6 are congruent, then line *m* is parallel to line *n*.

If angle 3 and angle 7 are congruent, then line *m* is parallel to line *n*.

If angle 4 and angle 8 are congruent, then line *m* is parallel to line *n*.

Way 3

If angle 1 and angle 8 are congruent, then line *m* is parallel to line *n*.

If angle 2 and angle 7 are congruent, then line *m* is parallel to line *n*.

BRAIN TICKLERS
Set # 12

Are these two lines parallel?

Answer each of the following questions *Yes, No,* or *Don't know.*

1. m∠1 = m∠5

2. m∠1 = m∠4

3. m∠1 = m∠3

4. m∠4 = m∠5

5. m∠1 = m∠8

(Answers are on page 72.)

Study Strategies

Draw a pair of parallel lines and a transversal on nine index cards. Using small arcs, draw one pair of angles on each card. Make sure to draw one card with a pair of vertical angles, adjacent angles, supplementary angles, complementary angles, interior angles, alternate interior angles, exterior angles, alternate exterior angles, and corresponding angles. Label the name of the angle on the back of each card. Use the cards to memorize the types of angles.

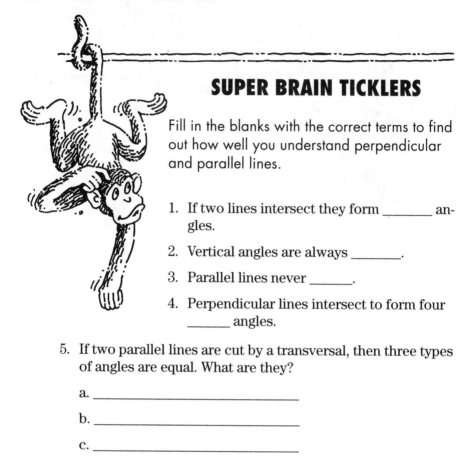

SUPER BRAIN TICKLERS

Fill in the blanks with the correct terms to find out how well you understand perpendicular and parallel lines.

1. If two lines intersect they form _____ angles.

2. Vertical angles are always _____.

3. Parallel lines never _____.

4. Perpendicular lines intersect to form four _____ angles.

5. If two parallel lines are cut by a transversal, then three types of angles are equal. What are they?

   a. _____

   b. _____

   c. _____

   d. _____

6. If two parallel lines are cut by a transversal, then any two angles are either congruent or _____.

7. Through a point not on a line there is (are) _____ perpendicular line(s) to the given line.

(Answers are on page 72.)

BRAIN TICKLERS—THE ANSWERS

Set # 9, page 55

1. 165
2. 15
3. 165
4. 180
5. 180
6. 180
7. 180
8. 360

Set # 10, page 59

1. False
2. False
3. False
4. True
5. True

Set # 11, page 63

1. 110
2. 70
3. 70
4. 110
5. 70
6. 70
7. 110

Set # 12, page 69

1. Yes

2. Don't Know

3. Don't Know

4. Yes

5. Yes

Super Brain Ticklers, page 70

1. four

2. congruent

3. intersect

4. right

5. a. Corresponding

   b. Alternate exterior

   c. Alternate interior

   d. Vertical angles

6. supplementary

7. one

CHAPTER FOUR

Triangles

INTERIOR ANGLES

A *polygon* is a closed figure with three or more sides. The sides of a polygon are line segments. Each side of a polygon intersects two other sides of the polygon at their endpoints. The endpoints of the sides of a polygon are the vertices of the polygon. A polygon lies in a single plane.

A *triangle* is the simplest type of polygon. A triangle has three sides and three angles. Each side of a triangle is labeled by its endpoints.

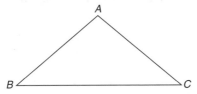

The sides of this triangle are $\overline{AB}$, $\overline{BC}$, and $\overline{CA}$.
When labeling the sides of a triangle, it doesn't matter which letter comes first.

> Side $\overline{AB}$ could also be labeled $\overline{BA}$.
> Side $\overline{BC}$ could also be labeled $\overline{CB}$.
> Side $\overline{CA}$ could also be labeled $\overline{AC}$.

Each two sides of a triangle form an angle. Where any two sides of a triangle meet is a *vertex* of the triangle. A triangle has three vertices.

A triangle has three *angles*, which are actually *interior angles*. Each of the interior angles of a triangle is less than 180 degrees.

The angles of a triangle are labeled three different ways.

1. The interior angles of a triangle can be labeled using three letters. The center letter is the vertex of the angle. The marked angle is $\angle ABC$. It could also be labeled $\angle CBA$.

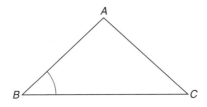

In this triangle, the marked angle is ∠*BAC* or ∠*CAB*.

2. The interior angle can be labeled by the single point that is the vertex of the angle, if and only if no other angles share the same vertex. The angles of this triangle can be labeled ∠*A*, ∠*B*, and ∠*C*.

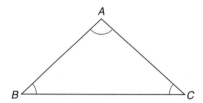

3. The interior angles can also be labeled by placing a small letter or number in the vertex of the angle. The angles of this triangle can be label ∠*a*, ∠*b*, and ∠*c*.

There are three ways to label the same angle. Look at triangle *XYZ*.

Angle *YXZ*, angle *X*, and angle 1 are all the same angle.
Angle *XYZ*, angle *Y*, and angle 2 are all the same angle.
Angle *XZY*, angle *Z*, and angle 3 are all the same angle.

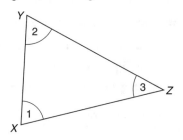

Experiment

Learn the types of angles that cannot form triangles.

Materials
  Protractor
  Pencil
  Paper

Procedure

  1. Draw a reflex angle. This is an angle greater than 180 degrees.
  2. Try to make a triangle using this angle. Be sure to use the reflex angle in the triangle.

Something to think about . . .
  Is there another type of angle that you cannot make a triangle out of?

Theorem: Every angle of a triangle has a measure greater than 0 and less than 180 degrees.

Experiment

Find the sum of the angles of a triangle.

Materials
    Paper
    Pencil
    Ruler
    Scissors

Procedure

1. Draw a triangle on a piece of paper.

2. Cut out the triangle.
3. Rip off the three angles of the triangle.

4. Put the three angles in a row so that the angles meet at one point and at least one side of each angle touches the side of another angle.

They form a straight line. The sum of the angles of a triangle is 180 degrees.

5. Draw a second triangle and repeat Steps 2 and 3.

Something to think about . . .

Can you draw a triangle where the sum of the angles is not 180 degrees?

Experiment

Measuring the angles of a triangle.

Materials

Protractor
Pencil
Paper

Procedure

1. Draw five triangles on a piece of paper. Make the triangles as different as you can.
2. Label the triangles One, Two, Three, Four, Five.
3. Using a protractor, measure the angles of each triangle. Enter the results in the chart.

4. Add the measurements of each of the angles of each triangle together. Put the results in the chart.

Triangles	Angle 1	Angle 2	Angle 3	Sum of Angles
One				
Two				
Three				
Four				
Five				

Did the sum of the angles of each of your triangles add up to 180 degrees?

Something to think about . . .
If you add the measures of the angles of a square, what is the sum? Does every square have the same number of degrees?

Theorem: The sum of the measures of the angles of a triangle is 180 degrees.

If you add the measures of the angles of any triangle together, the answer will always be 180 degrees.

If you know the measure of any two angles of a triangle, follow these two painless steps to compute the measure of the third angle.

Step 1: Add the measure of the two angles you know.

Step 2: Subtract this sum from 180 degrees. The result is the measure of the third angle.

EXAMPLE:

If a triangle has angles measuring 60 degrees and 40 degrees, what is the measure of the third angle?

Step 1: Add the two angles.

$$60 + 40 = 100$$

Step 2: Subtract the sum from 180.

$$180 - 100 = 80$$

The third angle is 80 degrees.

EXAMPLE:

Find the measure of a third angle of a right triangle that has an acute angle measuring 40 degrees.

Step 1: Add the two angles.
Since it is a right triangle, one angle is 90 degrees.

$$90 + 40 = 130$$

Step 2: Subtract the sum from 180.

$$180 - 130 = 50$$

The third angle is 50 degrees.

BRAIN TICKLERS
SET # 13

Find the measure of the missing angle in each of these triangles.

1. What is the measure of angle C in triangle ABC if the measure of angle A is 70 degrees and angle B is 80 degrees?

2. If the measure of angle A is 150 degrees and angle B is 20 degrees, what is the measure of angle C in triangle ABC?

3. If the measure of one acute angle of a right triangle is 10 degrees, what is the measure of the other acute angle?

(Answers are on page 110.)

EXTERIOR ANGLES

A triangle has six exterior angles. Each exterior angle of a triangle forms a straight line with an interior angle of the triangle. Extend one side of a triangle. The angle formed is an exterior angle.

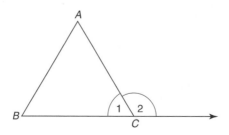

The interior angle is angle 1. Angle 2 is the exterior angle. Angle 1 is inside the triangle, while angle 2 is outside the triangle. Notice that the interior angle and the exterior angle form a straight line. The sum of angle 1 and angle 2 equals 180 degrees. Angle 1 and angle 2 are supplementary angles.

At each interior angle of a triangle, you can actually form *two* exterior angles. Extend both sides of the triangle at point *C* to form two exterior angles.

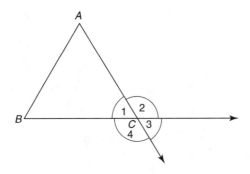

Angle 1 is the original interior angle.
Angle 2 is an exterior angle.
Angle 3 is a vertical angle formed by the two intersecting lines.
Angle 4 is the other exterior angle.

Notice the relationships between these four angles.

Angle 1 and angle 2 are supplementary. The sum of their measures equals 180 degrees.
Angle 1 and angle 4 are also supplementary. The sum of their measures equals 180 degrees.
Angle 1 and angle 3 are vertical angles. They are congruent.
Angle 2 and angle 4 are vertical angles. They are congruent.

Every triangle has six possible exterior angles. In this diagram all the exterior angles of one triangle are drawn. How many pairs of vertical angles can a triangle have?

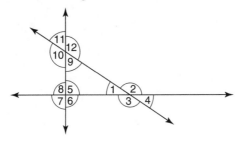

The exterior angles of this triangle are 2, 3, 6, 8, 10, and 12.

Experiment

Explore the exterior angles of a triangle.

Materials
　　Protractor
　　Pencil
　　Paper
　　Ruler

Procedure

1. Draw an acute triangle. Extend one of the sides of the triangle to form an exterior angle.

2. Label the exterior angle *a*. Label the interior angles 1, 2, and 3. In the triangle shown here,

   Angle *a* is the exterior angle.
   Angle 3 is the adjacent interior angle.
   Angles 1 and 2 are the nonadjacent interior angles.

3. Measure the exterior angle. Enter your answer in the chart.
4. Measure each of the nonadjacent interior angles. Enter your answers in the chart.
5. Add the measures of angles 1 and 2. Enter the result in the chart.

6. Draw a right triangle.

7. Extend one of the sides of the right triangle.

8. Label the exterior angle *a*. Label the interior angles 1, 2, and 3. In the above triangle,

Angle *a* is the exterior angle.
Angle 3 is the adjacent interior angle.
Angles 1 and 2 are nonadjacent interior angles.

9. Measure the exterior angle. Enter the answer in the chart.
10. Measure each of the nonadjacent interior angles. Enter the results in the chart.
11. Add the measures of angles 1 and 2. Enter the result in the chart.
12. Draw an obtuse triangle.

13. Extend one of the sides of the obtuse triangle.

14. Label the exterior angle *a*. Label the interior angles 1, 2, and 3.
    In the above triangle,

    Angle *a* is the exterior angle.
    Angle 3 is the adjacent interior angle.
    Angles 1 and 2 are the nonadjacent interior angles.

15. Measure the exterior angle. Enter your answer in the chart.
16. Measure each of the nonadjacent interior angles. Enter the results in the chart.
17. Add the measures of angles 1 and 2. Enter your answer in the chart.

Type of triangle	Measure of exterior angle a	Measure of adjacent interior angle 3	Measure of non-adjacent interior angle 1	Measure of non-adjacent interior angle 2	Sum of angles 1 and 2
Acute triangle					
Right triangle					
Obtuse triangle					

Do any two columns of the chart always match? Which ones?

Something to think about . . .
    What is the sum of all six exterior angles of a triangle? Is it the same for every triangle?

Theorem: The measure of an exterior angle of a triangle is equal to the sum of the measures of the two nonadjacent interior angles.

Theorem: The measure of any exterior angle of a triangle is greater than the measure of either of the two nonadjacent interior angles.

BRAIN TICKLERS
Set # 14

$m\angle 1 = 25$

$m\angle 2 = 60$

What is the measure of these angles?

1. Angle 3

2. Angle 4

3. Angle 5

4. Angle 6

5. Angle 7

6. Angle 8

7. Angle 9

8. Angle 10

9. Angle 11

10. Angle 12

(Answers are on page 110.)

TYPES OF TRIANGLES

Triangles are labeled by the size of their angles. There are acute, obtuse, and right triangles.

Acute triangles

An *acute* triangle is a triangle in which every angle has measures less than 90 degrees. Every angle of an acute triangle is an acute angle.

Here are three acute triangles.

Triangle 1   Triangle 2   Triangle 3

Experiment

Explore the angles of
the acute triangles.

Materials
   Protractor
   Pencil
   Ruler
   Paper

Procedure

   1. Measure the angles of the acute triangles in this picture.
   2. Enter the results in the chart.

Triangle 1   Triangle 2   Triangle 3

Triangle	Angle *a*	Angle *b*	Angle *c*
Triangle 1			
Triangle 2			
Triangle 3			

Something to think about . . .
   What did you notice about the angles of these three triangles?

BRAIN TICKLERS
Set # 15

Label each statement true or false.

1. A triangle with angles of 50, 60, and 70 degrees is an acute triangle.

2. The sum of the angles of an acute triangle is less than 180 degrees.

3. A triangle with angles of 10 and 20 degrees is an acute triangle.

(Answers are on page 111.)

Obtuse triangles

An *obtuse* triangle has one obtuse angle. An obtuse angle is an angle that is greater than 90 degrees.

Experiment

Explore obtuse triangles.

Materials
   Protractor
   Straight edge
   Pencil
   Paper

Procedure

1. Draw an obtuse angle.

2. Use one of the sides of this obtuse angle as the side of a second obtuse angle.
3. Try to draw a triangle using these two obtuse angles.

4. Draw two more obtuse angles. You can make the angles any size as long as they are greater than 90 degrees. Try to draw an obtuse triangle using these two angles. What happened?

Something to think about . . .
    What is always true about the measures of the other two
    angles of an obtuse triangle?

BRAIN TICKLERS
Set # 16

Label each statement true or false.

1. A triangle with angles of 10, 20, and 150 degrees is an obtuse triangle.

2. A triangle with two 40-degree angles is an obtuse triangle.

3. An obtuse triangle can have two obtuse angles.

(Answers are on page 111.)

Right triangles

A right triangle has one right angle. A right triangle has one angle that measures 90 degrees.

In this right triangle, angle *ABC* is a right angle. The other two angles are acute angles. A triangle can have at most one right angle.

Experiment

Explore right triangles.

Materials
    Protractor
    Pencil
    Ruler
    Paper

Procedure

1. Draw five different right triangles. The legs of the triangles should be the length shown in the chart.
2. Measure the angles of each right triangle using a protractor.
3. Look for patterns in the chart.

Triangle	Leg 1	Leg 2	Measure of angle A	Measure of angle B	Measure of angle C	Measure of hypotenuse
Triangle 1	1 inch	2 inches	90			
Triangle 2	3 inches	6 inches	90			
Triangle 3	5 inches	5 inches	90			
Triangle 4	1 inch	7 inches	90			
Triangle 5	2 inches	2 inches	90			

Something to think about . . .
    What is always true about the other two angles of a right triangle? (Hint: Find the sum of angle B and angle C.)

    What other relationships can you discover about right triangles?

A right triangle has three sides. The side opposite the right angle is called the hypotenuse of the triangle.

Theorem: The hypotenuse of a right triangle is longer than either two of the other sides.

BRAIN TICKLERS
Set # 17

Answer each of these statements true or false.

1. A right triangle can have two other angles that measure 60 and 60 degrees.

2. A triangle with angles measuring 30, 60, and 90 degrees is a right triangle.

3. A triangle with two 45-degree angles is a right triangle.

(Answers are on page 111.)

MORE TYPES OF TRIANGLES

Another way to classify triangles is to look at the number of congruent angles in a triangle. Some triangles have two congruent angles, some have three congruent angles, and in some triangles, none of the angles are congruent.

Isosceles triangles

An *isosceles triangle* has two congruent angles. An isosceles triangle has two congruent sides.

In this triangle, angle 1 and angle 2 are congruent. Angle 1 and angle 2 have the same measure. Sides $\overline{AB}$ and $\overline{AC}$ are congruent.

Theorem: The sides opposite the congruent angles of an isosceles triangle are congruent.

Experiment

Measure the angles of an isosceles triangle.

Materials
    Paper
    Pencil
    Protractor
    Scissors

Procedure

1. Fold a piece of paper in half.
2. Cut a triangle out of the piece of paper along the fold by cutting a "v" in the paper with one cut perpendicular to the fold. Open the folded piece of paper. It's an isosceles triangle.

3. Cut out three more isosceles triangles. Make them all different sizes and shapes. Measure the congruent angles of each isosceles triangle.

Triangle	Measure of congruent angles
Triangle 1	
Triangle 2	
Triangle 3	
Triangle 4	

Something to think about . . .

How many of the congruent angles are greater than 90 degrees?

How many of the congruent angles are less than 90 degrees?

How many of the congruent angles are equal to 90 degrees?

Theorem: The congruent angles of an isosceles triangle each measure less than 90 degrees.

BRAIN TICKLERS
Set # 18

Decide whether each of these statements is true or false.

1. A triangle with 30, 60, and 90-degree angles is an isosceles triangle.

2. An isosceles triangle can have two 90-degree angles.

3. A triangle with angles that measure 40 degrees and 100 degrees is an isosceles triangle.

(Answers are on page 111.)

Equiangular triangles

An *equiangular triangle* has three congruent angles.

Each angle of an equiangular triangle is 60 degrees. There are always 180 degrees in a triangle. If you divide 180 degrees by 3 the result is 60.

Equilateral triangles

A triangle with three equal sides is an *equilateral triangle*.

Experiment

Explore equiangular and equilateral triangles.

Materials
    Ruler
    Protractor
    Pencil
    Paper

Procedure

1. Draw an equiangular triangle. All three of the angles of the triangle should measure 60 degrees.
2. Measure the sides of the triangle. Are all three sides of the triangles equal? If a triangle has three equal sides, it is an equilateral triangle.
3. Draw a second equiangular triangle. Make sure all three angles of this triangle equal 60 degrees.
4. Measure the sides of this second triangle. Are all three sides of this second triangle equal?
5. Draw a triangle with three 4-inch sides.

6. Measure the angles of this triangle using a protractor.
   What is the measure of each of the angles of this triangle?

Something to think about . . .
   Is every equiangular triangle an equilateral triangle?
   Is every equilateral triangle an equiangular triangle?

BRAIN TICKLERS
Set # 19

Determine whether each of these statements is true or false.

1. A triangle with three 60-degree angles is an equiangular triangle.

2. A triangle with two 60-degree angles is an equiangular triangle.

3. A triangle with three 40-degree angles is an equiangular triangle.

(Answers are on page 111.)

Scalene triangles

A *scalene triangle* has no congruent sides and no congruent angles.

A scalene triangle can be an acute, obtuse, or a right triangle. A triangle whose angles measure 100 degrees, 45 degrees, and 35 degrees is an obtuse triangle because it contains an obtuse angle. It is also a scalene triangle since the measurement of all three angles is different.

Experiment

Explore angles and sides of scalene triangles.

Materials
    Paper
    Pencil
    Ruler

Procedure

1. Draw a scalene triangle. Label the angles of △ABC.
2. Measure the angles of the triangles.
3. Rank order the angles of the triangle from smallest to largest. Enter the results in the chart.
4. Measure the sides of the triangle.
5. Rank order the sides of the triangle from smallest to largest. Enter the results in the chart.

Name of angle	Rank of angle measure	Rank of side length	Name of side
Angle *A*			*AB*
Angle *B*			*BC*
Angle *C*			*AC*

6. Draw another scalene triangle. Repeat steps 1 to 5.

Name of angle	Rank of angle measure	Rank of side length	Name of side
Angle *A*			*AB*
Angle *B*			*BC*
Angle *C*			*AC*

Something to think about . . .
    Is there a relationship between the largest angle and the longest side?
    Is there a relationship between the smallest angle and the shortest side?

Theorem: The side of a triangle opposite the largest angle of a triangle is the largest side of the triangle.

BRAIN TICKLERS
Set # 20

Decide whether each of the following statements is true or false.

1. A triangle with angles measuring 50, 60, and 70 is a scalene triangle.

2. A triangle with angles measuring 170 and 2 degrees is a scalene triangle.

3. A triangle with angles measuring 45 and 90 is a scalene triangle.

4. A right triangle with a 30-degree angle is a scalene triangle.

(Answers are on page 112.)

BRAIN TICKLERS
Set # 21

Here are the measurements of three angles of a triangle. What type of triangles are these? Mark each set of measurements with the letters that apply.

A = Acute triangle
E = Equilateral triangle
I = Isosceles triangle
O = Obtuse triangle
R = Right triangle
S = Scalene triangle

1. Angles: 90 degrees, 45 degrees, 45 degrees

2. Angles: 80 degrees, 40 degrees, 60 degrees

3. Angles: 90 degrees, 60 degrees, 30 degrees

4. Angles: 60 degrees, 60 degrees, 60 degrees

5. Angles: 75 degrees, 75 degrees, 30 degrees

6. Angles: 30 degrees: 45 degrees, 105 degrees

7. Angles: 110 degrees, 35 degrees, 35 degrees

8. Angles: 10 degrees, 70 degrees, 100 degrees

(Answers are on page 112.)

PERIMETER OF A TRIANGLE

The *perimeter* of a triangle is the distance around the triangle. To find the perimeter of a triangle, add the length of each of the three sides of the triangle together.

Here is the painless solution for finding the perimeter of a triangle.

Step 1: Draw the triangle and label the sides.

Step 2: Add the measurements.

EXAMPLE:
What is the perimeter of a triangle with sides 4 centimeters,
5 centimeters, and 7 centimeters?

Step 1:

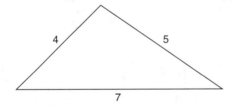

Step 2: Add the three sides of the triangle:

$$4 \text{ cm} + 5 \text{ cm} + 7 \text{ cm} = 16 \text{ cm}$$

The perimeter of the triangle is 16 centimeters.

EXAMPLE:
Find the perimeter of an equilateral triangle if each of the sides
of the triangle is 2 inches.

Step 1: Draw a triangle and label the sides.

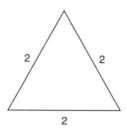

Step 2: Add the three sides: 2 in. + 2 in. + 2 in. = 6 in.

BRAIN TICKLERS
Set # 22

1. What is the perimeter of a triangle with sides 4, 10, and 16?

2. What is the perimeter of an equilateral triangle with side 3?

3. What is the perimeter of an isosceles right triangle with hypotenuse length 6 and sides length 2?

(Answers are on page 112.)

AREA OF A TRIANGLE

The altitude of a triangle is the line segment from the vertex of the triangle perpendicular to the opposite side.

Each triangle has three altitudes. The altitudes of this triangle are *AE*, *CD*, and *BF*.

Two of the altitudes of this obtuse triangle are on the exterior of the triangle.
The altitudes of this obtuse triangle are $\overline{AE}$, $\overline{BD}$, and $\overline{CF}$.

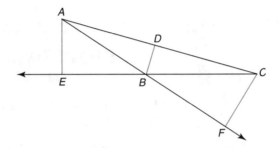

Two of the altitudes of a right triangle are the legs of the triangle.
AB is one of the altitudes of this right triangle.
BC is also an altitude of this right triangle.
BD is the final altitude of this right triangle.

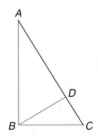

Experiment

See how the area of a triangle is related to the area of a rectangle.

Materials
    Scissors
    Graph paper
    Pencil
    Straight edge

Procedure

    1. On a piece of graph paper, draw a right triangle 8 squares long and three squares tall.

    2. Draw another right triangle with the same height and width.

    3. Cut out both triangles and fit them together.

    The two triangles should form a rectangle.

4. The area of a rectangle is base times height, which is written as $b \times h$. Notice that each of these triangles is exactly one half of the rectangle. Remember this and you'll never forget the area of a triangle is $\frac{1}{2}(b \times h)$ or $\frac{1}{2}$ base × altitude.

The area of the above rectangle is 24 square units.
The area of each of the triangles is $\frac{1}{2}(24)$ or 12 square units.

Theorem: The area of a triangle is $\frac{1}{2}$ base × altitude.

EXAMPLE:
Find the area of a triangle with base 10 and altitude 1.

The area of a triangle is $\frac{1}{2}$(base)(altitude).

The area is $\frac{1}{2}(10)(1)$.

The area is 5.

EXAMPLE:
Find the area of a right triangle with legs 7 and 10.
Use the length of one of the legs of the triangle as the base of a triangle and use the length of the other leg as the altitude of the triangle.

The area is $\frac{1}{2}(10)(7)$, which is $\frac{1}{2}(70) = 35$.

BRAIN TICKLERS
Set # 23

1. What is the area of the triangle with base 4 inches and altitude 6 inches?

2. What is the area of a right triangle with legs 5 and 8 inches long?

(Answers are on page 112.)

THE PYTHAGOREAN THEOREM

The square of the length of the hypotenuse of a right triangle is equal to the sum of the squares of the lengths of the legs.

Look at a right triangle. It has three sides. The side opposite the right angle is called the hypotenuse. The hypotenuse is a diagonal line that connects the other two sides of the triangle, called the legs.

Hypotenuse

Leg

Leg

Hypotenuse

Legs

If you have a right triangle, you can determine the length of any side if you know the length of the other two sides.

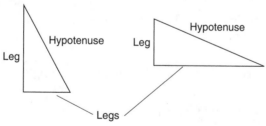

If you know the length of both legs of a right triangle, just follow these three painless steps to find the length of the hypotenuse.

Step 1: Square the length of the legs of the triangle.

Step 2: Add the squares of the legs.

Step 3: Take the square root of the result of Step 2. The result is the length of the hypotenuse of the triangle.

EXAMPLE:

If one leg of a right triangle is 3 inches long and the other leg of a right triangle is 4 inches long, how long is the hypotenuse?

Step 1: Square the length of legs. (Remember, to square a number, multiply the number by itself.)

Leg 1 is 3 inches. 3^2 is 9 square inches.
Leg 2 is 4 inches. 4^2 is 16 square inches.

Step 2: Add the squares of the legs.

$$9 + 16 = 25$$

Step 3: Take the square root of 25 square inches.

The square root of 25 square inches is 5 inches.
The length of the hypotenuse of this triangle is 5 inches.

If you know the length of one leg and the hypotenuse of a right triangle, just follow these four painless steps to find the length of the other leg of the triangle

Step 1: Square the length of the leg of the triangle.

Step 2: Square the length of the hypotenuse.

Step 3: Subtract the square of the leg of the triangle from the square of the hypotenuse.

Step 4: Find the square root of the result of Step 3.

EXAMPLE:

Find the length of a leg of a right triangle if one leg is 12 inches and the hypotenuse is 13 inches.

Step 1: Find the square of the leg.

$$12^2 = 12 \times 12 = 144$$

Step 2: Find the square of the hypotenuse.

$$13^2 = 13 \times 13 = 169$$

Step 3: Subtract the square of the leg from the square of the hypotenuse.

$$169 - 144 = 25$$

Step 4: Take the square root of the result of Step 3.

The square root of 25 is 5.
The length of the other leg is 5 inches.

The measurements of the sides of the triangle are 5, 12, and 13 inches.

BRAIN TICKLERS
Set # 24

1. If a right triangle has legs 2 inches long and 3 inches long, what is the square of the hypotenuse?

2. If a right triangle has sides 9 inches and 12 inches, what is the length of the hypotenuse?

3. If a triangle has one side of 3 inches and a 5-inch hypotenuse, what is the length of the third leg?

(Answers are on page 113.)

SUPER BRAIN TICKLERS

1. What are the measures of the angles of an isosceles right triangle?

2. What type of triangle has angles with measures 30-50-100?

3. What are the measures of the angles of an equilateral triangle?

4. What is the area of an isosceles right triangle with one leg 10 inches?

(Answers are on page 113.)

BRAIN TICKLERS—THE ANSWERS

Set # 13, page 82

1. 30 degrees

2. 10 degrees

3. 80 degrees

Set # 14, page 87

1. 95°

2. 155°

3. 25°

4. 155°

5. 85°

6. 95°

7. 85°

8. 120°

9. 120°

10. 60°

Set # 15, page 90

1. True

2. False

3. False

Set # 16, page 92

1. True

2. True

3. False

Set # 17, page 94

1. False

2. True

3. True

Set # 18, page 96

1. False

2. False

3. True

Set # 19, page 98

1. True

2. True

3. False

Set # 20, page 100

1. True

2. True

3. False

4. True

Set # 21, page 101

1. R, I

2. S, A

3. R, S

4. E, A

5. I, A

6. O, S

7. I, O

8. O, S

Set # 22, page 103

1. 30

2. 9

3. 10

Set # 23, page 107

1. 12 square inches

2. 20 square inches

Set # 24, page 109

1. 13 inches

2. 15 inches

3. 4 inches

Super Brain Ticklers, page 110

1. 45-45-90

2. Obtuse and scalene

3. 60-60-60

4. 50 square inches

CHAPTER FIVE

Similar and Congruent Triangles

TRIANGLE PARTS

Every triangle has three angles and three sides. There are special ways to refer to the angles and sides of a triangle and their relationship between each other. The words *adjacent, opposite,* and *included* are used to refer to the relationships between the sides and angles of a triangle.

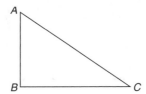

Sides $\overline{AB}$ and $\overline{AC}$ are both adjacent to angle A.
Sides $\overline{BC}$ and $\overline{BA}$ are both adjacent to angle B.
Side $\overline{CA}$ and $\overline{CB}$ are both adjacent to angle C.

Side $\overline{AB}$ is included between angles A and B.
Side $\overline{AC}$ is included between angles A and C.
Side $\overline{BC}$ is included between angles B and C.

Angle A is included between sides $\overline{AB}$ and $\overline{AC}$.
Angle B is included between sides $\overline{AB}$ and $\overline{BC}$.
Angle C is included between sides $\overline{BC}$ and $\overline{AC}$.

Side $\overline{BC}$ is opposite angle A.
Side $\overline{AB}$ is opposite angle C.
Side $\overline{AC}$ is opposite angle B.

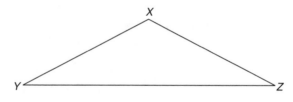

BRAIN TICKLERS
Set # 25

Look at triangle XYZ.

1. What angle is opposite side $\overline{XZ}$?

2. What side is opposite angle YZX?

3. What angle is included between sides $\overline{YX}$ and $\overline{XZ}$?

4. What side is included between angles Y and Z?

(Answers are on page 134.)

SIMILAR TRIANGLES

Look at these two triangles.

Triangle *ABC*

Triangle *DEF*

m∠*ABC* = m∠*DEF*
m∠*BCA* = m∠*EFD*
m∠*CAB* = m∠*FDE*

but

Side $\overline{AB}$ is *not equal* to side $\overline{DE}$.
Side $\overline{BC}$ is *not equal* to side $\overline{EF}$.
Side $\overline{CA}$ is *not equal* to side $\overline{FD}$.

These two triangles are not congruent, but they are similar.

- Similar triangles have equal angles, but they do not have equal sides.
- If two triangles are similar, one looks like a larger or smaller version of the other.
- If two triangles are similar, the sides of one are in direct proportion to the sides of the other.
- If two triangles are similar, the ratio of corresponding sides of one triangle is equal to the ratio of corresponding sides of another triangle.

Look at these two similar triangles.

If these two triangles are similar, then

$$\frac{AB}{XY} = \frac{BC}{YZ} = \frac{AC}{XZ}$$

You can use the ratios to find the length of missing sides of a triangle.

EXAMPLE:

Look at the following triangles.

If $AB = 2$
$BC = 3$
$AC = 5$
$DE = 8$

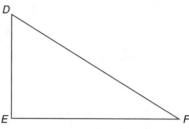

Find EF and DF.

First match up corresponding sides.

$$\frac{AB}{DE} = \frac{BC}{EF} = \frac{AC}{DF}$$

Now substitute the sides you know.

$$\frac{2}{8} = \frac{3}{EF} = \frac{5}{DF}$$

Take one pair of ratios.

$$\frac{2}{8} = \frac{3}{EF}$$

Cross-multiply to solve.

$$2(EF) = 3(8)$$
$$2(EF) = 24$$
$$EF = 12$$

Now solve for *DF*.

$$\frac{2}{8} = \frac{5}{DF}$$

$$2(DF) = 5(8)$$

$$2(DF) = 40$$

$$DF = 20$$

BRAIN TICKLERS
Set # 26

If *AB* = 2, *AC* = 1, *DE* = 5, and *CD* = 6, find *BC* and *CE*.

(Answers are on page 134.)

CONGRUENT TRIANGLES

Congruent triangles have the same size and the same shape. Two triangles are congruent if their angles are congruent *and* their sides are congruent.

These two triangles are congruent.

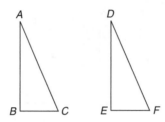

Angle $A \cong$ Angle D
Angle $B \cong$ Angle E
Angle $C \cong$ Angle F

Side $\overline{AB} \cong$ Side $\overline{DE}$
Side $\overline{BC} \cong$ Side $\overline{EF}$
Side $\overline{CA} \cong$ Side $\overline{FD}$

In order for two triangles to be congruent, all six of these statements must be true.

ALL TRIANGLES
ARE NOT
CREATED EQUAL

> **If two triangles are congruent, then their corresponding parts are equal.**

EXAMPLE:

If triangle *ABC* is congruent to triangle *DEF* and the measure of angle *ABC* is 90 degrees, then the measure of angle *DEF* is also 90 degrees.

If triangle *ABC* is congruent to triangle *DEF* and side *AB* = 4, then side *DE* = 4.

BRAIN TICKLERS
Set # 27

Triangle *ABC* is congruent to triangle *XYZ*.

The measure of angle *B* = 90, the measure of angle *A* = 20, and side *XY* = 2.

1. What is the measure of angle *X*?

2. What is the measure of angle *Y*?

3. What is the measure of angle *Z*?

4. What is the length of side *AB*?

(Answers are on page 134.)

Experiment

Discover the corresponding parts of congruent triangles.

Materials
>   Scissors
>   Glue
>   Pencil
>   Paper
>   Colored paper

Procedure

1. Fold a piece of white paper in half, and cut out five pairs of congruent triangles of different sizes and shapes. Since the paper is folded, every time you cut out a triangle you will get two triangles.
2. Place the matching pairs of triangles on the colored paper. Turn them upside down and over so that no two triangles look alike. Glue them in place.
3. Now find the congruent triangles and mark the corresponding angles and corresponding sides.

Something to thing about . . .
>   What is the easiest way to match corresponding parts?

SSS Postulate
If three sides of one triangle are congruent to three sides of another triangle, then the triangles are congruent.

EXAMPLE:

To prove triangle $ABC \cong$ triangle XYZ, show that $AB \cong XY$, $AC \cong XZ$, and $BC \cong YZ$.

Mini-Proof

Figure *ABDC* is a rectangle.
Show that triangle *ABD* is congruent to triangle *ACD*.

To prove these two triangles congruent, show that all three sides are equal to each other. In other words, $AC \cong BD$, $AB \cong CD$, and $AD \cong AD$.

First list what you know.

1. Figure *ABDC* is a rectangle.

2. Segment *AD* is a diagonal of rectangle *ABDC*.

Next list what you can infer based on what you know.

3. $\overline{AC} \cong \overline{BD}$ since they are opposite sides of a rectangle, and opposite sides of a rectangle are always congruent.

4. $\overline{AB} \cong \overline{CD}$ since they are opposite sides of a rectangle, and opposite sides of a rectangle are always congruent.

5. $\overline{AD} \cong \overline{AD}$ since every line segment is congruent to itself.

Conclusion
Since all three sides of one triangle are congruent to all three sides of the other triangle, the triangles are congruent. Since the triangles are congruent,

$$\angle ABD \cong \angle ACD$$
$$\angle DAB \cong \angle ADC$$
$$\angle BDA \cong \angle CAD$$

Experiment

Discover what happens when only two sides of one triangle are congruent to two sides of another triangle.

Materials
    Paper
    Pencil
    Ruler

Procedure

1. Draw a line segment 1 inch long. Draw another line segment 2 inches long that starts at one end of the 1-inch-long line segment. Connect the two line segments to form a triangle.
2. Draw another 1-inch-long line segment with another 2-inch-long line segment connected to it. Connect these two line segments to form another triangle.
3. How many different triangles can you draw using a 1-inch line segment and a 2-inch line segment?

Something to think about . . .
    How many different triangles can you find with two matching sides?
    Is it possible to prove two triangles equal with just two equal sides?

SAS Postulate
If two sides and the included angle of one triangle are congruent to two sides and the included angle of a second triangle, then the triangles are congruent.

EXAMPLE:

To prove triangle ABC = triangle XYZ, show that $AB \cong XY$, $AC \cong XZ$, and angle A is congruent to angle X. Angle A is included between sides AB and AC, and angle X is included between sides $\overline{XY}$ and $\overline{XZ}$. You can also prove that triangle ABC is congruent to triangle XYZ using angle B and angle Y and their adjacent sides, or angles C and Z and their adjacent sides.

Mini-Proof

Figure $WXZY$ is a rhombus.
$\overline{WZ}$ and $\overline{XY}$ are diagonals of the rhombus. Is triangle WEY congruent to triangle ZEX?

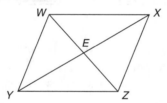

To prove triangles WEY and ZEX congruent using SAS, show that two sides and an included angle of one triangle are congruent to two sides and an included angle of another triangle.

First list what you know.

1. Figure $WXZY$ is a rhombus.

2. $\overline{WZ}$ is a diagonal of the rhombus.

3. $\overline{XY}$ is a diagonal of the rhombus.

Next list what you can infer based on what you know.

4. The measure of angle WEY = the measure of angle XEZ since they are vertical angles, and all vertical angles are equal. Since the measures are equal, the angles are congruent.

5. $\overline{WZ}$ bisects $\overline{XY}$ since the diagonals of a rhombus bisect each other.

6. $\overline{WE} = \overline{EZ}$ since a bisected line segment is divided into two congruent segments.

7. $\overline{XE} = \overline{EY}$ since a bisected line segment is divided into two congruent segments.

Conclusion
Triangle *WEY* is congruent to triangle *ZEX* since two sides and the included angle of one triangle are congruent to two sides and the included angles of another triangle.

ASA Postulate
If two angles and the included side of one triangle are congruent to two angles and the included side of another triangle, then the two triangles are congruent.

EXAMPLE:
To prove triangle *ABC* is congruent to triangle *XYZ* you must show that angle *A* is congruent to angle *X*, angle *B* is congruent to angle *Y*, and the included side $\overline{AB}$ is congruent to the included side $\overline{XY}$.

Mini-Proof

Figure *ABCD* is a square. $\overline{AD}$ and $\overline{BC}$ are diagonals of the square. Prove triangle *ADC* is congruent to triangle *BCD*.

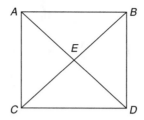

First list what you know.

1. *ABDC* is a square.

2. $\overline{AD}$ is a diagonal of square *ABDC*.

3. $\overline{BC}$ is a diagonal of square *ABDC*.

Next list what you can infer from what you know.

4. $\overline{CD} = \overline{CD}$ since every segment is congruent to itself.

5. Angle *C* ≅ angle *D* since they are both right angles.

6. $\overline{BC} = \overline{AD}$ since the diagonals of a square are congruent.

7. $\overline{EC} = \overline{ED}$ since the diagonals of a square bisect each other.

8. △ *ECD* is an isosceles triangle.

9. ∠*ECD* ≅ ∠*EDC*, since the base angles of an isosceles triangle are congruent.

Conclusion
Using the ASA postulate, triangle *ADC* and triangle *BCD* are congruent since two angles and the included side of each triangle are congruent. You could also prove these two triangles congruent using SAS.

AAS Postulate
Two triangles are congruent if two angles and the side opposite one of them are congruent.

EXAMPLE:

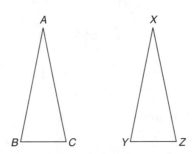

To prove triangle $ABC \cong$ triangle XYZ:

- show that angle A is congruent to angle X and angle B is congruent to angle Y. Now show that side $\overline{BC} \cong$ side $\overline{YZ}$ or side $\overline{AC} \cong$ side $\overline{XZ}$, since these are all nonincluded sides.

or

- show that angle B is congruent to angle Y and angle C is congruent to angle Z and either $\overline{AB} \cong \overline{XY}$ or $\overline{AC} \cong \overline{XZ}$, since these are all nonincluded sides.

or

- show that angle A is congruent to angle X and angle C is congruent to angle Z and $\overline{AB} \cong \overline{XY}$ or $\overline{BC} \cong \overline{YZ}$, since these are all nonincluded sides.

Mini-Proof

Prove that triangle ACD is congruent to triangle DBA.

$ABDC$ is a parallelogram. A parallelogram is a four-sided figure in which opposite sides are parallel and equal. $\overline{AD}$ is perpendicular to $\overline{CD}$ and $\overline{AB}$.

First list what you know.

1. $ABDC$ is a parallelogram.

2. $\overline{AD}$ is perpendicular to $\overline{CD}$ and $\overline{AB}$.

Next list what you can infer from what you know.

3. Angle B is congruent to angle C, since the opposite angles of a parallelogram are congruent.

4. $\overline{AD} \cong \overline{AD}$, since every segment is congruent to itself.

5. Angle ADC and angle DAB are right angles.

6. The measure of angle ADC = the measure of angle DAB, since all right angles are 90 degrees.

Conclusion
Triangle ACD is congruent to triangle DBA, since two angles and a nonincluded side of one triangle are equal to two angles and a nonincluded side of another triangle.

Caution—Major Mistake Territory!

You *cannot* prove two triangles congruent if two sides of one triangle are congruent to two sides of another triangle and an angle that is not included is congruent to an angle that is not included. *The angle must be included for the two triangles to be congruent.*

Hypotenuse-Leg Postulate
Two right triangles are congruent if the hypotenuse and leg of one right triangle are congruent to the hypotenuse and leg of another right triangle.

To prove that these two right triangles are congruent, just show that hypotenuse $\overline{AC}$ is congruent to hypotenuse $\overline{XZ}$ and leg $\overline{AB}$ is equal to $\overline{XY}$. Or show that hypotenuse $\overline{AC}$ is equal to hypotenuse $\overline{XZ}$ and leg $\overline{BC}$ is equal to $\overline{YZ}$.

Mini-Proof

Triangle ABC is an isosceles triangle where side $\overline{AB} \cong$ side $\overline{AC}$. Segment $\overline{AD}$ is the perpendicular bisector of side $\overline{BC}$. Prove triangle ABD is congruent to triangle ACD.

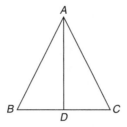

First list what you know.

1. Triangle *ABC* is an isosceles triangle.

2. $\overline{AD}$ is the altitude of triangle *ABC*.

   Next list what you can infer from what you know.

3. $\overline{AD}$ is perpendicular to $\overline{BC}$ since $\overline{AD}$ is an altitude of triangle *ABC*.

4. Angle *ADB* and angle *ADC* are both right angles since $\overline{AD}$ is perpendicular to $\overline{BC}$.

5. Triangle *ADB* and triangle *ADC* are both right triangles, since angle *ADB* and angle *ADC* are both right angles.

6. The hypotenuse of triangle *ABD* is congruent to the hypotenuse of triangle *ACD*, since triangle *ABC* is an isosceles triangle.

7. One leg of triangle *ABD* is congruent to one leg of triangle *ACD*, since $\overline{AD}$ is a leg of both triangles and congruent to itself.

Conclusion
Triangle *ABD* is congruent to triangle *ACD* because of the hypotenuse-leg postulate. The hypotenuse and one leg of triangle *ABD* is congruent to the hypotenuse and one leg of triangle *ACD*.

Study Strategies

There are five ways to prove triangles congruent.
Write each of these five ways on one side of an index card.

SSS   Side, side, side
SAS   Side, angle, side
ASA   Angle, side, angle
AAS   Angle, angle, side
H-L   Hypotenuse-leg

BRAIN TICKLERS—THE ANSWERS

Set # 25, page 118

1. Angle *Y*, which is also angle *XYZ*

2. Side *XY*

3. Angle *X* or angle *YXZ*

4. Side *YZ*

Set # 26, page 121

Side $BC = 2\frac{2}{5}$ and side $CE = 2\frac{1}{2}$

Set # 27, page 123

1. 20

2. 90

3. 70

4. 2

CHAPTER SIX

Quadrilaterals

Quadrilaterals are four-sided polygons. There are all kinds of quadrilaterals. Squares and rectangles are common quadrilaterals. Trapezoids, parallelograms, and rhombuses are also quadrilaterals. Quadrilaterals are everywhere. A playing card, a window frame, and index cards are all quadrilaterals.

TRAPEZOIDS

A trapezoid is a quadrilateral with exactly one pair of parallel sides. Both of these figures are trapezoids.

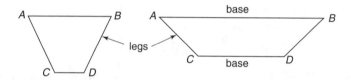

In both figures, side $\overline{AB}$ is parallel to side $\overline{CD}$. Notice that $\overline{AC}$ is not parallel to $\overline{BD}$.

The parallel sides of the trapezoid are called the *bases* of the trapezoid. $\overline{AB}$ and $\overline{CD}$ are the bases of these trapezoids.

The nonparallel sides of the trapezoid are called the *legs* of the trapezoid. $\overline{AC}$ and $\overline{BD}$ are the legs of these trapezoids.

The *median* of a trapezoid is a line segment that is parallel to both bases of the trapezoid and connects both legs of the trapezoid at their midpoints. In trapezoid $ABCD$, $\overline{EF}$ is the median of the triangle. The length of the median of a trapezoid is

$$\frac{1}{2} \text{ (base 1 + base 2).}$$

EXAMPLE:
If one base of a trapezoid is 6, and the other leg is 12 inches, what is the length of the median?

$$\text{Length of median} = \frac{1}{2} \, (6 + 12)$$

$$\text{Length of median} = \frac{1}{2} \, (18)$$

$$\text{Length of median} = 9$$

Here is a picture of a trapezoid.

What do you know about this trapezoid?

$\overline{AB}$ and $\overline{CD}$ are the bases of the trapezoid.
$\overline{AD}$ and $\overline{BC}$ are the legs of the trapezoid.
$\overline{AB}$, $\overline{CD}$, and $\overline{EF}$ are all parallel to each other.
$\overline{AE}$ is congruent to $\overline{ED}$.
$\overline{BF}$ is congruent to $\overline{FC}$.
$\overline{EF}$ is the median of the trapezoid.

We know many things about trapezoids. Do this experiment to find some common properties of trapezoids.

Experiment

Explore the properties of trapezoids.

Materials
    Paper
    Pencil
    Scissors
    Tape

Procedure

1. On a piece of paper, draw three large triangles. Draw one obtuse triangle, one right triangle, and one isosceles triangle.

2. Cut out all three of these triangles.
3. Draw a line parallel to the base of each of the triangles.

4. Cut off the top of the triangles on the line. The result is three trapezoids.
5. Rip off the angles on the same side of the trapezoid. Tape these angles together. Together they should equal 180 degrees.

6. Rip off all four angles from one of the trapezoids. Tape all four of the angles together. What is the sum of all four angles?

Something to think about . . .

If you draw a line on a trapezoid parallel to the base of the trapezoid, two new shapes result. What are they?

If you draw a line down the center of a trapezoid perpendicular to the base of the trapezoid, two new shapes result. What are they?

Theorem: Two angles on one side of a trapezoid are supplementary.

Angle A and angle C are supplementary.

$$m\angle A + m\angle C = 180$$

Angle B and angle D are supplementary.

$$m\angle B + m\angle D = 180$$

Theorem: The sum of the angles of a trapezoid is 360 degrees.

$$m\angle A + m\angle B + m\angle C + m\angle D = 360$$

Mini-Proof

Prove that the sum of the angles of a trapezoid is 360 degrees.

First list what you know.

1. *ABDC* is a trapezoid.

Next list what you can infer from what you know.

2. Angles *A* and *C* are supplementary since both angles on the same side of a parallelogram are supplementary.

3. Angles *B* and *D* are supplementary since both angles on the same side of a parallelogram are supplementary.

4. $m\angle A + m\angle C = 180$ since supplementary angles total 180 degrees.

5. $m\angle B + m\angle D = 180$ since supplementary angles total 180 degrees.

Conclusion

$$m\angle A + m\angle B + m\angle C + m\angle D = 360 \text{ degrees}$$

ISOSCELES TRAPEZOIDS

An *isosceles trapezoid* is a special type of trapezoid, where both legs are equal. To draw an isosceles trapezoid, start by drawing an isosceles triangle. Draw a line parallel to the base of the triangle. Cut off the top of the triangle along the line you just drew. The result is an isosceles trapezoid.

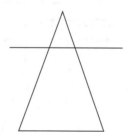

An isosceles trapezoid has two pairs of base angles.

Angles *A* and *B* are one pair of base angles of the trapezoid.

Angles *C* and *D* are the other pair of base angles of the trapezoid.

The base angles of an isosceles trapezoid are congruent.

Angle *A* and angle *B* are congruent.
Angle *C* and angle *D* are congruent.

The legs of an isosceles triangle are congruent.

$\overline{AC}$ is congruent to $\overline{BD}$.

The diagonals of an isosceles trapezoid are congruent.

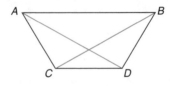

$\overline{AD}$ is congruent to $\overline{BC}$.

Experiment

Discover the sum of the angles of an isosceles trapezoid.

Materials
   Paper
   Pencil

Procedure

   1. Draw a trapezoid.
   2. Draw a single diagonal in the trapezoid. This diagonal divides the trapezoid into two triangles.

3. Label the triangles triangle 1 and triangle 2.
4. What is the measure of the sum of the angles of triangle 1?
5. What is the measure of the sum of the angles of triangle 2?
6. Add the sum of the angles of these two triangles together.

$$180 + 180 = 360$$

The total is 360 degrees.

Something to think about . . .

Is there another way to divide an isosceles trapezoid into two equal shapes?

The sum of the angles of an isosceles trapezoid is 360 degrees.

$$m\angle A + m\angle B + m\angle C + m\angle D = 360 \text{ degrees}$$

Theorem: The opposite angles of an isosceles trapezoid are supplementary.

EXAMPLE:

Angles 1 and 4 are supplementary.

$$m\angle 1 + m\angle 4 = 180 \text{ degrees}$$

Angles 2 and 3 are supplementary.

$$m\angle 2 + m\angle 3 = 180 \text{ degrees}$$

BRAIN TICKLERS
Set # 28

Compute the values of the angles of this isosceles trapezoid.

1. If m∠2 = 30 degrees, what is the measure of angle 5?

2. If m∠1 = 30 degrees, what is the measure of angle 4?

Compute the measures of the angles for the isosceles trapezoid if m∠3 is 120 degrees.

3. What is the measure of angle 1?

4. What is the measure of angle 2?

5. What is the measure of angle 4?

6. What is the measure of angle 5?

7. What is the measure of angle 6?

(Answers are on page 163.)

PARALLELOGRAMS

A *parallelogram* is a quadrilateral with two pairs of parallel sides. All these figures are parallelograms.

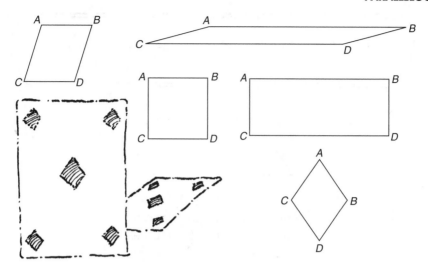

In all these figures, side AB is parallel to side CD, and side AC is parallel to side BD.

Theorem: The opposite sides of a parallelogram are congruent.

EXAMPLE:

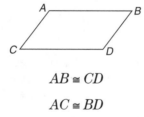

$$AB \cong CD$$
$$AC \cong BD$$

EXAMPLE:

Theorem: Opposite angles of a parallelogram are congruent.

Angle $A \cong$ angle D

Angle $C \cong$ angle B

145

Theorem: Consecutive pairs of angles of a parallelogram are supplementary.

EXAMPLE:

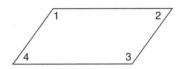

$m\angle 1 + m\angle 2 = 180$ degrees
$m\angle 2 + m\angle 3 = 180$ degrees
$m\angle 3 + m\angle 4 = 180$ degrees
$m\angle 4 + m\angle 1 = 180$ degrees

Experiment

Explore the relationship between the angles of a parallelogram.

Materials
   Protractor
   Pencil
   Paper

Procedure

1. Draw three different parallelograms.
2. Label the angles of each parallelogram angles 1, 2, 3, and 4.

3. Measure each of the angles of the parallelogram. Enter the measurements in the chart.

Parallel-ogram	$m\angle 1$	$m\angle 2$	$m\angle 3$	$m\angle 4$	$m\angle 1+$ $m\angle 2$	$m\angle 1+$ $m\angle 4$	$m\angle 2+$ $m\angle 3$	$m\angle 3+$ $m\angle 4$	$m\angle 1+m\angle 2+$ $m\angle 3+m\angle 4$
1									
2									
3									

4. Add the measurements of the pairs of angles indicated in the chart. Enter the results in the chart.

5. Add the measurements of all four angles together. Enter the results in the chart.

Something to think about . . .

What did you notice about the patterns formed?

Can you draw a parallelogram where the sum of the angles is not 360 degrees?

Theorem: The sum of the angles of a parallelogram is 360 degrees.

EXAMPLE:

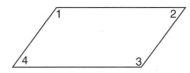

$$m\angle 1 + m\angle 2 + m\angle 3 + m\angle 4 = 360$$

Theorem: Each diagonal of a parallelogram separates the parallelogram into two congruent triangles.

EXAMPLE:

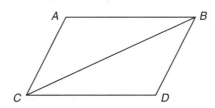

Triangle $ABC \cong$ Triangle BCD

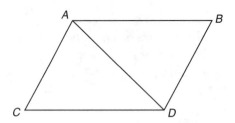

Triangle *ABD* ≅ Triangle *ACD*

Mini-Proof

How can you prove triangle *ABC* is congruent to triangle *CAD*?

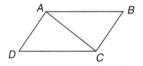

First list what you know.

1. *ABCD* is a parallelogram.

2. $\overline{AC}$ is a diagonal of parallelogram *ABCD*.

Now list what you can infer from what you know.

3. $\overline{AC} \cong \overline{AC}$ since every segment is congruent to itself.

4. $\overline{AB} \cong \overline{CD}$ since opposite sides of a parallelogram are congruent.

5. $\overline{AD} \cong \overline{BC}$ since opposite sides of a parallelogram are congruent.

Conclusion
Triangle *ABC* is congruent to triangle *CAD* since three sides of one triangle are congruent to three sides of another triangle.

Theorem: The diagonals of a parallelogram bisect each other.

EXAMPLE:

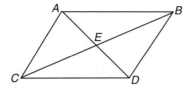

Diagonals AD and BC intersect at point E.

$$\overline{AE} = \overline{ED}$$

$$\overline{CE} = \overline{EB}$$

Experiment

Compare the diagonals of a parallelogram.

Materials
    Pencil
    Paper
    Scissors

Procedure

1. Draw a parallelogram. Connect points A and D. Call this parallelogram 1.
2. Copy the parallelogram. Connect points B and C. Call this parallelogram 2.
3. Cut parallelogram 1 on diagonal AD.
4. Cut parallelogram 2 on diagonal BC.
5. Compare the length of diagonal AD to diagonal BC. Are they both the same length?

Something to think about . . .
    Can you draw a parallelogram where the diagonals are equal? What shape is it?

BRAIN TICKLERS
Set # 29

In this parallelogram the measure of angle 5 is 110 degrees.

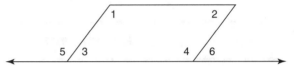

1. What is m∠1?

2. What is m∠2?

3. What is m∠3?

4. What is m∠4?

5. What is m∠6?

Find the measure of the indicated angles when m∠5 = 30, m∠6 = 40, and m∠7 = 50.

6. What is m∠1?

7. What is m∠2?

8. What is m∠8?

9. What is m∠11?

10. What is m∠12?

(Answers are on page 163.)

RHOMBUSES

A *rhombus* is a parallelogram with four equal sides. Since it is a parallelogram, opposite sides of a rhombus are parallel. All these shapes are rhombuses.

> **Theorem:** The diagonals of a rhombus bisect the angles of the rhombus.

EXAMPLE:

Diagonal $\overline{AD}$ bisects angle *CAB* and angle *BDC*.

>   The measure of angle 1 is equal to the measure of angle 2.
>   The measure of angle 7 is equal to the measure of angle 8.

Diagonal $\overline{BC}$ bisects angle *ABD* and angle *DCA*.

>   The measure of angle 3 is equal to the measure of angle 4.
>   The measure of angle 5 is equal to the measure of angle 6.

> **Theorem:** The diagonals of a rhombus are perpendicular.

EXAMPLE:

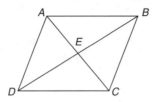

These diagonals intersect at point E.

$\overline{AC}$ is perpendicular to $\overline{BD}$.
$\angle AEB$ is a right angle.
$\angle BEC$ is a right angle.
$\angle CED$ is a right angle.
$\angle DEA$ is a right angle.

Since a rhombus is a special parallelogram with four equal sides, everything that applies to parallelograms also applies to rhombuses.

Theorem: The opposite angles of a rhombus are congruent.

EXAMPLE:

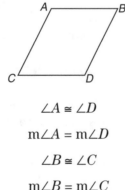

$$\angle A \cong \angle D$$

$$m\angle A = m\angle D$$

$$\angle B \cong \angle C$$

$$m\angle B = m\angle C$$

Theorem: The consecutive pairs of angles of a rhombus are supplementary.

EXAMPLE:

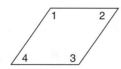

$$m\angle 1 + m\angle 2 = 180 \text{ degrees}$$

$$m\angle 2 + m\angle 3 = 180 \text{ degrees}$$

$$m\angle 3 + m\angle 4 = 180 \text{ degrees}$$

$$m\angle 4 + m\angle 1 = 180 \text{ degrees}$$

Theorem: The diagonals of a rhombus separate the rhombus into two congruent triangles.

EXAMPLE:

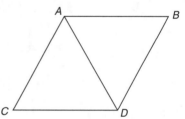

Triangle ABC is congruent to triangle BCD. Triangle ACD is congruent to triangle ABD.

Theorem: The diagonals of a rhombus bisect each other.

EXAMPLE:

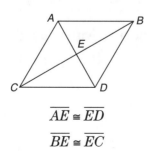

$$\overline{AE} \cong \overline{ED}$$

$$\overline{BE} \cong \overline{EC}$$

BRAIN TICKLERS
Set # 30

This figure is a rhombus. The measure of angle 5 is 40. Determine the measures of the indicated angles.

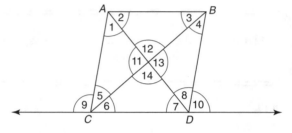

1. Angle *ACD*

2. Angle 4

3. Angle 1

4. Angle 9

5. Angle 11

(Answers are on page 164.)

RECTANGLES

A rectangle is a parallelogram with four right angles. Both figures are rectangles.

$$m\angle 1 = 90$$
$$m\angle 2 = 90$$
$$m\angle 3 = 90$$
$$m\angle 4 = 90$$

Theorem: All four angles of a rectangle are equal.

EXAMPLE:

Angle A = Angle B = Angle C = Angle D

Theorem: Any two angles of a rectangle are supplementary.

EXAMPLE:

$m\angle 1 + m\angle 2 = 180$ degrees; angle 1 and angle 2 are supplementary.

$m\angle 1 + m\angle 3 = 180$ degrees; angle 1 and angle 3 are supplementary.

$m\angle 1 + m\angle 4 = 180$ degrees; angle 1 and angle 4 are supplementary.

$m\angle 2 + m\angle 3 = 180$ degrees; angle 2 and angle 3 are supplementary.

$m\angle 2 + m\angle 4 = 180$ degrees; angle 2 and angle 4 are supplementary.

$m\angle 3 + m\angle 4 = 180$ degrees; angle 3 and angle 4 are supplementary.

Since a rectangle is a parallelogram, the theorems that apply to parallelograms also apply to rectangles.

> **Theorem:** The diagonals of a rectangle separate the rectangle into two congruent triangles.

EXAMPLE:

 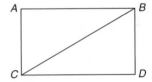

Triangle *ACD* is equal to triangle *ABD*.
Triangle *ABC* is equal to triangle *BCD*.

> **Theorem:** The diagonals of a rectangle bisect each other.

EXAMPLE:

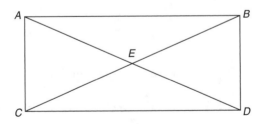

$$\overline{AE} \cong \overline{ED}$$
$$\overline{BE} \cong \overline{EC}$$

BRAIN TICKLERS
Set # 31

This figure is a rectangle, and m∠1 = 30.

1. What is the measure of angle 2?

2. What is the measure of angle 3?

3. What is the measure of angle 4?

4. What is the measure of angle 5?

5. What is the measure of angle 6?

(Answers are on page 164.)

SQUARES

A *square* is a rectangle with four equal sides.

```
A ┌────┼────┐ B
  │ □       □ │
  ┤         ├
  │         │
  │ □       □ │
C └────┼────┘ D
```

Since a square is a type of rectangle, the measure of all the angles of a square is 90 degrees.

$$m\angle A = 90$$

$$m\angle B = 90$$

$$m\angle C = 90$$

$$m\angle D = 90$$

> **Theorem:** The measures of all the angles of a square are equal.

EXAMPLE:

$$m\angle A = m\angle B = m\angle C = m\angle D$$

> **Theorem:** The sum of the angles of a square is 360 degrees.

EXAMPLE:

$$90 + 90 + 90 + 90 = 360$$

Since a square is a rectangle and a rectangle is a type of parallelogram, opposite sides of a square are parallel.

> **Theorem:** All four sides of a square are equal.

EXAMPLE:

$$\text{Side } \overline{AB} = \text{Side } \overline{BC} = \text{Side } \overline{CD} = \text{Side } \overline{DA}$$

Look at the relationship between the diagonals of a square.

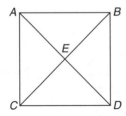

> **Theorem:** The diagonals of a square are equal.

EXAMPLE:

$$\text{Diagonal } \overline{AD} = \text{Diagonal } \overline{BC}$$

Theorem: The diagonals of a square bisect each other.

EXAMPLE:
Diagonal AD bisects diagonal BC.

$$\overline{AE} \cong \overline{ED}$$

$$\overline{BE} \cong \overline{EC}$$

Since the diagonals of a square are equal, and they are bisected, then

$$\overline{AE} = \overline{ED} = \overline{BE} = \overline{EC}$$

Theorem: The diagonals of a square are perpendicular to each other.

EXAMPLE:
Diagonal $\overline{AD}$ is perpendicular to diagonal $\overline{BC}$.

Angle AEC is a right angle.
Angle AEB is a right angle.
Angle BED is a right angle.
Angle CED is a right angle.

Theorem: The triangles formed by the diagonals of a square are congruent.

EXAMPLE:

Triangle ABD = Triangle ACD = Triangle ABC = Triangle BCD

Caution—Major Mistake Territory!

Every square is a rectangle, but every rectangle is not a square.

Every square is a parallelogram, but every parallelogram is not a square.

Every square is a quadrilateral, but every quadrilateral is not a square.

Every rectangle is a parallelogram, but every parallelogram is not a rectangle.

Every rectangle is a quadrilateral, but every quadrilateral is not a rectangle.

Every parallelogram is a quadrilateral, but every quadrilateral is not a parallelogram.

Experiment

Discover the relationship between the four small triangles formed by the diagonals of a square.

Materials
  Paper
  Pencils
  Scissors

Procedure

1. Draw a square.
2. Draw the diagonals of the square.
3. Cut the square along the diagonals and form four small triangles.
4. Place these four triangles on top of each other. Are they identical?
5. What other shapes can you construct from these four small triangles?

Something to think about . . .
Do the diagonals of a rectangle form four identical triangles?

BRAIN TICKLERS
Set # 32

Determine the relationship between the following pairs of angles.

C = Complementary angles
S = Supplementary angles
E = Equal angles
V = Vertical angles
A = Adjacent angles

1. Angle *CAB* and angle *BDC*

2. Angle 1 and angle 2

3. Angle 1 and angle 4

4. Angle 3 and angle 6

5. Angle 1 and angle 5

(Answers are on page 164.)

Study Strategies

Fill in the following summary chart.
Check the boxes that are true for each quadrilateral.

Features	Trapezoid	Parallelogram	Rhombus	Rectangle	Square
Opposite angles are congruent.					
Opposite angles are supplementary.					
All angles are congruent.					
Diagonals bisect each other.					
Diagonals are perpendicular.					
Opposite sides are congruent.					
Adjacent angles are supplementary.					
Sum of the angles is 360.					

SUPER BRAIN TICKLERS

Answer each of the following questions by circling what applies.

Q represents Quadrilateral
T represents Trapezoid
P represents Parallelogram
RH represents Rhombus
R represents Rectangle
S represents Square

Q	T	P	RH	R	S	1. All four sides are parallel.
Q	T	P	RH	R	S	2. All four angles are right angles.
Q	T	P	RH	R	S	3. Diagonals are perpendicular.
Q	T	P	RH	R	S	4. Only two sides are parallel.
Q	T	P	RH	R	S	5. All the angles can be different.
Q	T	P	RH	R	S	6. Diagonals are equal.
Q	T	P	RH	R	S	7. Opposite angles are congruent.
Q	T	P	RH	R	S	8. Opposite angles are supplementary.

(Answers are on page 165.)

BRAIN TICKLERS—THE ANSWERS

Set # 28, page 144

1. 30

2. 150

3. 60

4. 60

5. 120

6. 60

7. 60

Set # 29, page 150

1. 110°

2. 70°

3. 70°

4. 110°

5. 70°

6. 30°

7. 40°

8. 60°

9. 90°

10. 90°

Set # 30, page 154

1. 80 degrees

2. 40 degrees

3. 50 degrees

4. 100 degrees

5. 90 degrees

Set # 31, page 157

1. 90

2. 60

3. 30

4. 90

5. 60

Set # 32, page 161

1. S, E

2. A, C, E

3. E, C

4. S, E

5. C, E

Super Brain Ticklers, page 163

1. P, RH, R, S

2. R, S

3. RH, S

4. T

5. Q, T

6. R, S

7. P, RH, R, S

8. R, S

CHAPTER SEVEN

Circles

Circles are everywhere. A wheel is in the shape of a circle. So are the face of a clock, the rim of a plate, and the lip of a cup. Circles are unique geometric shapes. Every circle has three things in common.

1. Every circle has a center. The center is a fixed point in the middle of the circle.

2. Every point the exact same distance from the center of the circle is on the circle.

3. All the points in a circle are on a single plane.

Circles are usually named by their center. This is circle Q.

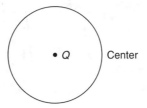

Center

Experiment

Learn a creative way to make a circle.

Materials
    Piece of string 5-inches long
    Two pencils
    Paper

Procedure

    1. Tie one end of the piece of string tightly around a pencil.
    2. Tie the other end of the string around a second pencil.
    3. Place one of the pencils, point down, firmly in the center of a piece of paper.
    4. Keeping the string tight, draw a circle by moving the second pencil around the first.
    5. Use a different length of string and draw a different size circle with the same center.

Something to think about . . .
How does a compass work to draw circles?
What else could you use to draw a circle?

Concentric circles are circles with the same center.

RADIUS AND DIAMETER

The *radius* of a circle is a line segment that starts at the center of the circle and ends on the circle. Every radius of the same circle is exactly the same length.

EXAMPLE:

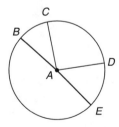

$\overline{AB}$, $\overline{AC}$, $\overline{AD}$, and $\overline{AE}$ are all radii of the circle.
All these radii are the same length.
$\overline{AB} = \overline{AC} = \overline{AD} = \overline{AE}$

The *diameter* of a circle is a line segment that connects both sides of the circle and passes through the center of the circle. All diameters of a single circle are the same length.

EXAMPLE:

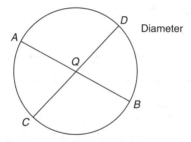

$\overline{AB}$ and $\overline{CD}$ are both diameters of circle Q.
$\overline{AB} = \overline{CD}$

Experiment

Discover the relationship between the radius and the diameter of a circle.

Materials
   Compass
   Paper
   Pencil
   Ruler

Procedure

1. Use a compass to draw a large circle.
2. Use the compass to draw a smaller circle.
3. Draw three lines through the center of each circle. These are the diameters of each circle.
4. Measure the length of each diameter. Enter the results in the chart.
5. Draw three radii in each circle.
6. Measure the length of each radius. Enter the results in the chart.

Radius/diameter	Small circle	Large circle
Radius 1		
Radius 2		
Radius 3		
Diameter 1		
Diameter 2		
Diameter 3		

Something to think about . . .
Are all the radii from one circle the same length?
Are all the diameters from one circle the same length?
What is the relationship between the length of the radius and the length of the diameter?

Theorem: Two circles are congruent if and only if their diameters are the same length.

Theorem: Two circles are congruent if and only if their radii are the same length.

You can prove that two circles are congruent if you can prove that either their radii or diameters are congruent.

BRAIN TICKLERS
Set # 33

1. If a circle has a 3-inch radius, what is the length of the diameter?

2. If a circle has a 12-inch diameter, what is the length of the radius?

(Answers are on page 197.)

REGIONS OF A CIRCLE

A circle divides a plane into three distinct areas:

- interior of a circle
- exterior of a circle
- the circle itself.

The area inside the circle is called the *interior of the circle*. Points *A*, *B*, and *C* lie in the interior of the circle.

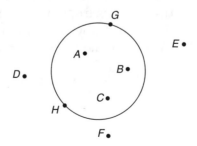

Points G and *H* lie on the circle. The exterior of the circle is all the points that are farther from the center of the circle than the length of the radius of the circle. Points *D*, *E*, and *F* lie in the exterior of a circle.

The distance from the center of the circle to any point inside the circle is less than the length of the radius. The distance from the center of the circle to any point in the exterior of the circle is greater than the length of the radius. The distance from the center of the circle to any point on the circle is exactly the length of the radius.

BRAIN TICKLERS
Set # 34

1. A circle has a radius of 4 inches. A point lies six inches from the center of the circle. Is this point on the circle, in the interior of the circle, or in the exterior of the circle?

2. A circle has a diameter of 10 inches. A point lies eight inches from the center of the circle. Is this point on the circle, in the interior of the circle, or in the exterior of the circle?

3. A circle has a diameter of 12 inches. A point lies six inches from the center of the circle. Is this point in the interior of the circle, in the exterior of the circle, or on the circle?

4. A circle has a radius of 5 inches. A point lies three inches from the center of the circle. Is this point in the interior of the circle, in the exterior of the circle, or on the circle?

(Answers are on page 197.)

PI

Pi is a Greek symbol and is written as the symbol π. Pi is determined by dividing the circumference of a circle by its diameter. Pi is $3.141592\ldots$ and continues forever. Pi is a nonterminating, nonrepeating decimal. When you solve problems where you need the number pi, use an approximation of pi. The number 3.14 is a common approximation of pi. So is $\frac{22}{7}$.

Experiment

Determine the relationship between the circumference and the diameter of a circle.

Materials
　　String
　　Ruler
　　Pencil
　　Paper
　　Calculator
　　Plate, saucer, cup, glass, garbage pail, quarter, and other round items

Procedure

1. Measure the distance around each round object by putting a string around each object. Measure the length of the string used to circle each item using the ruler. Write your findings in the chart.
2. Measure the length of the diameter using the ruler and the string. Make sure that the diameter goes through the center of the circle. Write your findings in the chart.
3. Divide the distance around each circle (circumference) by the distance across the center of the circle (diameter). Write your results in the chart.
4. Compare the results of column 3 for the glass, cup, saucer, quarter, plate, and garbage pail.

Round item	Length of circumference	Length of diameter	Length of circumference divided by length of diameter
Glass			
Cup			
Saucer			
Quarter			
Plate			
Round garbage pail			

Something to think about . . .

If the diameter of one circle is twice as large as another circle, what is the relationship between the circumferences of these two circles?

BRAIN TICKLERS
Set # 35

Determine whether each of the following statements is true or false.

1. Pi is a Roman symbol.
2. Pi is approximately 3.14.
3. Pi is the same as π.

(Answers are on page 198.)

CIRCUMFERENCE OF A CIRCLE

The distance around a circle is called the circumference and is measured in linear units.

To figure out the circumference of a circle, follow these painless steps.

Step 1: Find the diameter of the circle.

Step 2: Multiply the diameter by pi.

Circumference of a Circle

$$C = d(\pi)$$

d stands for the diameter of a circle.
To approximate the circumference of a circle, use either of these formulas:

$$C = d\left(\frac{22}{7}\right) \quad \text{or} \quad C = d(3.14)$$

EXAMPLE:

Compute the circumference of a circle with diameter 4.

$$C = 3.14d = 3.14(4) = 12.56$$

Computing the circumference of a circle using the radius is painless.

Step 1: Find the radius of the circle.

Step 2: Multiply the radius by 2 since the diameter is twice the radius.

Step 3: Multiply the diameter by pi.

> **Circumference of a Circle**
> $$C = 2r(\text{pi})$$

EXAMPLE:

Compute the circumference of a circle with radius 1.

Step 1: Find the radius of the circle.

The radius is 1.

Step 2: Multiply the radius by 2.

(2)(1) = 2
The diameter is 2.

Step 3: Multiply the diameter by π.

2(3.14) = 6.28
The circumference of the circle is 6.28.

A *semicircle* is half a circle. The endpoints of a semicircle are a diameter of a circle.

To find the perimeter of a semicircle, follow these four painless steps.

Step 1: Find the circumference of the entire circle.

Step 2: Divide the circumference by 2 to find the length of the portion of the circle that is the semicircle.

Step 3: Find the length of the diameter.

Step 4: Add the results of Step 2 and Step 3.

EXAMPLE:
Find the circumference of a semicircle with diameter 10.

10

Step 1: Find the circumference of the entire circle.
The circumference of the entire circle is 10(3.14) or 31.4.

Step 2: Divide the circumference by 2 to find the length of the portion of the circle that is the semicircle.

$$\frac{31.4}{2} = 15.7$$

Step 3: Find the length of the diameter.
The length of the diameter is 10 units.

Step 4: Add the results of Step 2 and Step 3. Add the length of the arc (15.7) to the length of the diameter (10) to find the circumference of the semicircle.

$$15.7 + 10 = 25.7$$

Theorem: Congruent circles have the same circumference.

BRAIN TICKLERS
Set # 36

1. Find the circumference of a circle with diameter 10.

2. Find the circumference of a circle with radius 5.

3. Find the circumference of a circle with radius $\frac{1}{2}$.

4. Find the circumference of a semicircle with radius 3.

5. Find the radius of a circle with circumference 6(pi).

(Answers are on page 198.)

AREA OF A CIRCLE

The *area* of a circle is the area covered by the interior of the circle and the circle itself.

Experiment

Learn to approximate the area of a circle.

Materials
　　Graph paper
　　Pencil
　　Compass

Procedure

1. Draw a circle on a piece of graph paper.

2. Count the number of complete squares inside the circle. Put an x on each square you count to make sure that you don't count the same square twice. Write this number down.
3. Count the number of partial squares inside the circle. Shade each of these partial squares. Multiply the number of partial squares by $\frac{1}{2}$. Write this number down.
4. Add the numbers in Step 2 and Step 3.

Something to think about . . .
　　Is there an easier way to find the area of a circle?

Finding the area of a circle is painless.

Step 1: Find the radius of the circle.

Step 2: Multiply pi times the radius squared.

Area of a Circle
$$A = \text{pi}(r)^2$$

EXAMPLE:
Find the area of a circle with radius 3.

Area of a circle = pi(3)²

Area of a circle = 9 pi square units

You can also find the area of a circle if you know the diameter.

Area of a Circle
$$A = \text{pi}\left(\frac{d}{2}\right)^2$$

EXAMPLE:
What is the area of a circle with a diameter of 6 inches?

First find the radius of the circle.
The radius of the circle is half of the diameter.
$\frac{1}{2}(6) = 3$ inches
Area of a circle = pi(3)²
Area of a circle = 9 pi square inches

You can also find the area of a circle if you know the circumference. Just follow these painless steps.

Step 1: First divide the circumference by pi to find the diameter.

Step 2: Divide the diameter by 2 to find the radius of the circle.

Step 3: Square the radius of the circle.

Step 4: Multiply the result of Step 3 by pi.

EXAMPLE:

Find the area of a circle with circumference 10 pi.

Step 1: First divide the circumference by pi to find the diameter.

> 10 pi divided by pi is 10.
> The diameter of the circle is 10.

Step 2: Divide the diameter by 2 to find the radius of the circle.

> 10 divided by 2 is 5.
> The radius of the circle is 5.

Step 3: Square the radius of the circle.

$$5^2 = 25$$

Step 4: Multiply the result of Step 3 by pi.

$$25 \text{ pi}$$

The area of the circle is 25 pi square units.

BRAIN TICKLERS
Set # 37

1. What is the area of a circle with radius 10?

2. What is the area of a circle with diameter 2?

3. What is the area of a circle with circumference 20 pi?

(Answers are on page 198.)

Experiment

Compare the area of a circle with the area of a square.

Materials
    Graph paper
    Pencil
    Calculator

Procedure

1. Draw a square with a side of 10 units on the graph paper.
2. Inscribe a circle inside the square. To inscribe a circle in a square, each side of the square should be touched by the circle once. This circle has a diameter of 10 units.

3. Compute the area of the square. (Area = s × s)
4. Computer the area of the circle. (Area = 3.14 × r × r)
5. Which area is larger?

6. Divide the area of the circle by the area of the square. Multiply the answer by 100 to change the answer to a percent.
7. What percent of the area of the square is the area of the circle?
8. Repeat the first seven steps for a circle with a diameter of 2 units and a square with side 2.

Something to think about . . .

How can the results be used to estimate the area of a circle? Inscribe a square in a circle. What percentage of the area of a circle is the area of a square?

Theorem: Congruent circles have the same area. If two circles have the same area, they must be congruent.

To prove two circles congruent, show that

- their radii are the same length
- their diameters are the same length
- their circumferences are the same length
- their areas are the same

DEGREES IN A CIRCLE

A right angle has a measure of 90 degrees. The measure of the angles in a triangle is 180 degrees. How many degrees in a circle?

The degrees of a circle can be determined by measuring the central angles of a circle. A *central angle* is an angle whose vertex is the center of the circle and whose sides intersect the sides of a circle.

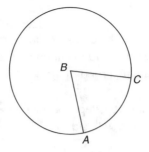

Angle *ABC* is a central angle. Use a protractor to measure a central angle.

Experiment

Discover the number of degrees in a circle.

Materials
    Paper
    Pencil
    Protractor

Procedure

1. Measure each of the central angles in these circles using a protractor. Enter the results in the chart.
2. Add the measures of the angles. Enter your answer in the chart.

Circle 1

Circle 2

Angle	Circle 1	Circle 2
Angle 1		
Angle 2		
Angle 3		
Angle 4		
Angle 5		
Total of angles 1, 2, 3, 4, and 5		

Something to think about . . .

Draw a small circle approximately half the size of Circles 1 and 2. How many degrees are in this circle?

Draw a large circle approximately twice the size of Circles 1 and 2. How many degrees are in this circle?

Theorem: Every circle has a total of 360 degrees.

BRAIN TICKLERS
Set # 38

1. How many degrees are in a circle with radius 4?

2. How many degrees are in a circle with diameter 10?

3. How many degrees are in a circle with radius 100?

(Answers are on page 198.)

CHORDS AND TANGENTS

A *chord* is a line segment whose endpoints lie on the circle. Segments $\overline{AB}$, $\overline{CD}$, $\overline{EF}$, and $\overline{GH}$ are all chords.

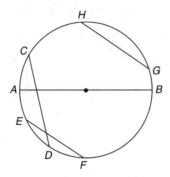

Notice that segment $\overline{AB}$ is both a chord and a diameter of the circle.

These line segments are not chords.

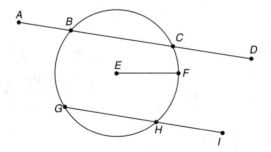

Two of these segments ($\overline{AD}$ and $\overline{GI}$) extend beyond the circle. One of these segments ($\overline{EF}$) is the radius of the circle. It contacts only one side of the circle. However, $\overline{BC}$ and $\overline{GH}$ are chords.

A *tangent* is a line on the exterior of the circle that touches the circle in exactly one point.

EXAMPLE:

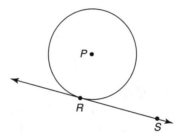

Line *RS* is tangent to circle *P*.
The circle and the line intersect at exactly one point, *R*.

BRAIN TICKLERS
Set # 39

Decide whether each of the following statements is true or false.

1. A radius is a chord.

2. A tangent intersects a circle at two points.

3. A diameter is a chord.

(Answers are on page 198.)

ARCS

An *arc* is a portion of a circle.

A *semicircle* is an arc that is exactly half a circle. The endpoints of a semicircle lie on the diameter of a circle.

EXAMPLE:

Draw a circle. Draw a diameter and call it $\overline{AB}$.

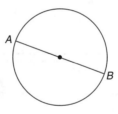

Each semicircle is called $\overset{\frown}{AB}$.

Not all arcs are semicircles.

EXAMPLE:

Draw a circle and pick two points on the circle. These two points divide the circle into two arcs. One of the arcs is smaller than the other arc.

The smaller arc is called the *minor arc*. A minor arc is less than a semicircle. Label the minor arc using two letters. Place an arc sign over the letters. Arc AB is written as $\overset{\frown}{AB}$.

The larger arc is called a *major arc*. A major arc is larger than a semicircle. Label the major arc using three letters by picking another letter on the arc. Three letters are used to label major arcs so they are not confused with the minor arcs. Place an arc sign over the three letters. Arc *ACB* is written as $\overset{\frown}{ACB}$.

BRAIN TICKLERS
Set # 40

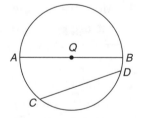

Determine whether each of the following statements is true or false. Segment *AB* is a diameter. Segment *CD* is a chord that does not go through the center of the circle.

1. Arc *AB* is a semicircle.

2. Arc *CD* is a major arc.

3. Arc *CAB* is greater than a semicircle.

(Answers are on page 199.)

MEASURING ARCS

The measure of an arc is expressed in degrees, while the length of an arc is expressed in linear measurements such as inches, centimeters, and feet.

Finding the measure of an arc is easy. Just measure the central angle that the arc cuts off. Remember that a central angle is an angle that has its vertex as the center of the circle and its sides are radii of the circle.

To find the measure of an arc, just follow these painless steps.

Step 1: Draw a line segment from one end of the arc to the center of the circle.

Step 2: Draw a line segment from the other end of the arc to the center of the circle.

Step 3: Measure the angle created by the two line segments. If the arc is a minor arc (less than 180 degrees), this is the measure of the arc.

Step 4: If the arc is a major arc (greater than 180 degrees) subtract the measure of the angle from 360 degrees to find the measure of the major arc.

EXAMPLE:

Find the measure of the arc indicated.

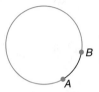

Step 1: Connect one end of the arc to the center of the circle.

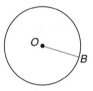

Step 2: Connect the other end of the arc to the center of the circle.

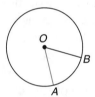

Step 3: Measure the angle created by the two line segments.

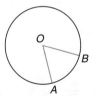

Angle *AOB* is 60 degrees.

Step 4: If the arc is a major arc (greater than 180 degrees), subtract the measure of the angle from 360 degrees to find the measure of the major arc. Since the arc is a major arc, subtract 60 degrees from 360 degrees.

$$360 - 60 = 300$$

The measure of the arc is 300 degrees.

To find the length of an arc, just follow these painless steps.

Step 1: Find the circumference of the circle.

Step 2: Divide the measure of the arc in degrees by 360.

Step 3: Multiply the result of Step 1 by the result of Step 2.

EXAMPLE:

Find the length of an arc with a measure of 30 degrees that is part of a circle with radius 12 inches.

Step 1: Find the circumference of the circle.
The circumference of the circle = pi(diameter).
If the radius is 12, the diameter is 24.
The circumference is 24 pi.

Step 2: Divide the measure of the arc in degrees by 360.
The arc is 30 degrees. So, 30 divided by 360 is $\frac{1}{12}$.

Step 3: Multiply the result of Step 1 by the result of Step 2.
Multiply 24 pi $\left(\frac{1}{12}\right)$.
The answer is 2pi or 6.28.

Theorem: Congruent arcs have the same degree measure and the same length.

Arc *AB* and arc *CD* have both the same measure and the same length.

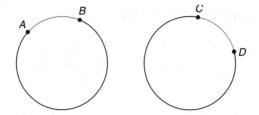

Arc *EF* and arc *GH* are not congruent. They have the same length but not the same measure.

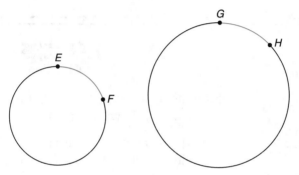

These two arcs are not congruent. They have the same measure but not the same length.

Caution—Major Mistake Territory!

Congruent arcs must come from congruent circles. Just because two arcs contain the same number of degrees does not mean that they are congruent.

BRAIN TICKLERS
Set # 41

1. Find the measure of an arc that is 180 degrees in a circle with diameter 4.

2. Find the measure of an arc that is 270 degrees in a circle with diameter 10.

3. Find the measure of an arc that is 45 degrees in a circle with diameter 2.

(Answers are on page 199.)

Study Strategies

Write "Area of a Circle" on an index card. On the other side of the card write pi(r)².
Write "Circumference of a Circle" on an index card. On the other side of the card write pi(d) or 2 pi(r). Write the rest of the formulas and definitions in this chapter on index cards. Study them.

SUPER BRAIN TICKLERS

1. What is the area of a circle with diameter 12?

2. What is the circumference of a circle with radius 4?

3. What is the circumference of a circle with area 9 pi?

4. What is the area of a circle with circumference 10 pi?

5. What is the length of a 60-degree arc on a circle with radius 12?

(Answers are on page 199.)

BRAIN TICKLERS—THE ANSWERS

Set # 33, page 173

1. 6 inches
2. 6 inches

Set # 34, page 174

1. Exterior of the circle
2. Exterior of the circle
3. On the circle
4. Interior of the circle

Set # 35, page 177

1. False
2. True
3. True

Set # 36, page 180

1. 31.4
2. 31.4
3. 3.14
4. 15.42
5. 3

Set # 37, page 183

1. 314 square units
2. 3.14 square units
3. 314 square units

Set # 38, page 187

1. 360 degrees
2. 360 degrees
3. 360 degrees

Set # 39, page 189

1. False
2. False
3. True

Set # 40, page 192

1. True
2. False
3. True

Set # 41, page 196

1. 2 pi
2. 7.5 pi
3. 0.25 pi

Super Brain Ticklers, page 197

1. 36 pi
2. 8 pi
3. 6 pi
4. 25 pi
5. 4 pi

CHAPTER EIGHT

Perimeter, Area, and Volume

IT'S A MATTER OF UNITS

It's possible to measure the perimeter, area, surface area, and volume of various shapes. The *perimeter* is the distance around a shape. The perimeter is measured in inches, feet, yards, miles, centimeters, meters, and kilometers.

The *area* is the amount of flat space a flat shape encloses. Square units are used to measure area. Examples of square units are square inches, square feet, square yards, square miles, square centimeters, square meters, and square kilometers. Write square units by putting a small two over the units to indicate "square."

> Square inches = in.^{2}
> Square feet = ft^2
> Square miles = mi^2
> Square meters = m^2

Surface area is the outside surface of a solid shape. Surface area is measured in square units.

Volume is the space inside a three-dimensional shape. A liter of water and a cup of sugar are measured in volume. Volume is measured in cubic units. Think of each cubic unit as a little block or cube. Write cubic units by putting a small three over the units.

> Cubic inches = in.^{3}
> Cubic feet = ft^3
> Cubic miles = mi^3
> Cubic centimeters = cm^3
> Cubic meters = m^3
> Cubic kilometers = km^3

It's not hard to learn how to find the perimeter, area, surface area, and volume of common figures.

TRIANGLES

A triangle has three sides and three angles.

Perimeter

The perimeter of a triangle is the distance around a triangle. The perimeter of a triangle is the sum of the sides of a triangle.

> Perimeter of a triangle = Side 1 + Side 2 + Side 3

EXAMPLE:
Find the perimeter of a triangle with sides 8, 10, and 12 feet.

Just add the three sides together.
8 + 10 + 12 = 30 feet

Area

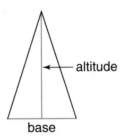

The area of a triangle is the space inside the triangle. The area of a triangle is $\frac{1}{2}$ base × altitude. The *altitude* is the distance from a vertex of the triangle perpendicular to the opposite side. The altitude is also called the *height*.

$$\text{Area of a triangle} = \frac{1}{2}ba$$

$$\text{Area of a triangle} = \frac{1}{2}bh$$

EXAMPLE:

What is the area of a triangle with base 8 and altitude 10?

The area is $\frac{1}{2}(8)(10)$.

The area is 40.

BRAIN TICKLERS
Set # 42

1. Find the perimeter of an equilateral triangle with sides of 5 inches each.

2. Find the perimeter of an isosceles triangle with two sides of 4 inches each and a third side 1 inch long.

3. What's the area of a triangle with base 4 inches and height 1 inch?

(Answers are on page 247.)

RECTANGLES

A *rectangle* is a parallelogram. It has four right angles and two pairs of parallel sides.

Perimeter

To find the perimeter of a rectangle, just add the length of all four sides together.

> Perimeter of a rectangle = $\ell + \ell + w + w$

EXAMPLE:
Find the perimeter of a rectangle with sides 5 inches and 7 inches.

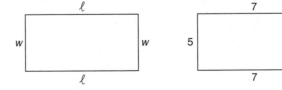

Just add all four sides of the rectangle together.
7 + 7 + 5 + 5 = 24
The perimeter is 24 inches.

Experiment

Learn how to determine the area of a rectangle by counting squares.

Materials
　　Graph paper
　　Pencil

Procedure

1. Draw the rectangles indicated on the chart on a piece of graph paper.
2. Count the number of small squares inside each drawn rectangle. Enter the results in the chart.
3. Compute the area of each rectangle by multiplying the length by the width of each rectangle. Enter the results in the chart.
4. Compare the areas you found in Steps 2 and 3.

Rectangle	Length	Width	Area found by counting	Area found by multiplying $l \times w$
Rectangle 1	6	5		
Rectangle 2	10	3		
Rectangle 3	2	4		
Rectangle 4	4	2		
Rectangle 5	1	1		

Something to think about . . .
   How could you determine the area of a rectangle with sides
   10 inches × 12 inches?

Area

To determine the area of a rectangle, just multiply the length
times the width.

$$\text{Area of a rectangle} = l \times w$$

EXAMPLE:
To find the area of a rectangle 2 inches by 4 inches, just multiply
2 × 4.
   The area of this rectangle is 8 square inches.

BRAIN TICKLERS
Set # 43

1. What is the perimeter of a rectangle with length 10 inches and height 6 inches?

2. What is the perimeter of a rectangle with sides 4 meters and 16 meters?

3. What is the area of a rectangle with length 10 inches and height 5 inches?

4. What is the area of a rectangle with length 4 centimeters and height 2 centimeters?

(Answers are on page 247.)

SQUARES

A *square* has four right angles and four equal sides. A square is a special type of rectangle. Since all four sides of a square are equal, the length and width of the square are called the sides of the square.

Perimeter

To find the perimeter of a square, just add all the sides together.

Perimeter of a Square = $s + s + s + s$

EXAMPLE:

What is the perimeter of a square with sides 9 inches long?

Add all four sides of the square together.
9 + 9 + 9 + 9 = 36

Experiment

Discover the formula for finding the perimeter of a square.

Materials
    Pencil
    Paper

Procedure

1. Draw each of the squares listed in the chart.
2. Label each of the sides.
3. Add the sides of each square together to find its perimeter. Enter the results in the chart.
4. Multiply the length of one side of each square by the number of sides (4). Enter the results in the chart.

	Length of one side, s	Perimeter found by adding the sides	Perimeter found by multiplying 4 times s
Square 1	5		
Square 2	7		
Square 3	1		
Square 4	10		

5. Did you get the same result each way? Which way was the easiest?

Something to think about . . .

Can you find a shortcut method to find the perimeter of a rectangle?

Perimeter of a square = $s + s + s + s$

Perimeter of a square = $4(s)$,   which is 4 times s

Area

The area of a square is length times width. Since the length and width of a square are exactly the same, the area of a square is $s \times s$ or s^2.

Area of a square = $s \times s$

Experiment

Compute the area of different shapes.

Area = MESS

Materials
    Scissors
    Pencil
    Magic marker
    Paper

Procedure

    1. Draw a square on a piece of paper.
    2. Draw both diagonals of the square with magic marker.
    3. Cut out the square.
    4. Cut the square along the diagonals.

    5. Create a new shape with these triangles. Trace the new
       shape on a piece of paper.
    6. Next create another new shape. Trace this shape.
    7. All these shapes have the same area.

Something to think about . . .
    Is the perimeter of each shape you constructed the same?

BRAIN TICKLERS
Set # 44

1. What is the perimeter of a square with side 7?

2. What is the area of a square with side 5?

3. What is the area of a square with perimeter 12?

4. What is the perimeter of a square with area 100 square units?

(Answers are on page 247.)

PARALLELOGRAMS

A *parallelogram* is a quadrilateral with two pair of parallel sides. All squares and rectangles are parallelograms, but not all parallelograms are squares and/or rectangles.

Perimeter

To find the perimeter of a parallelogram, add the length of all four sides.

Perimeter of a parallelogram = $l + w + l + w$

EXAMPLE:

Find the perimeter of this parallelogram.

$$6 + 10 + 6 + 10 = 32$$

Area

To find the area of a parallelogram multiply length times the altitude. The altitude is a line segment drawn from one side of the parallelogram perpendicular to the opposite side.

Area of a parallelogram = $l(a)$

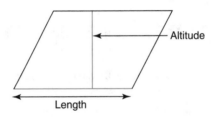

EXAMPLE:

What is the area of a parallelogram with length 8, width 5, and altitude 4?

Multiply the length (8) by the altitude (4).
8 × 4 = 32
The area is 32 square units.

Caution—Major Mistake Territory!

Don't compute the area of a parallelogram by multiplying the length times the width. You must multiply the altitude times the length.

Experiment

Change a parallelogram to a rectangle.

Materials
　　Pencil
　　Graph paper
　　Scissors
　　Tape

Procedure

1. Draw a parallelogram on a piece of graph paper. The parallelogram you draw should *not* be a rectangle.
2. Make the parallelogram 7 squares long and 4 squares tall.
3. Cut out the parallelogram.
4. Cut a triangle off one end of the parallelogram. Start at the inside corner and cut straight down. Look at the diagram.

5. Slide the triangle to the right and tape it to the other end of the parallelogram. The parallelogram is now a rectangle.

6. What is the area of the rectangle created? Multiply length times width.

7. What is the area of the original parallelogram? Multiply length times altitude.

8. How do the two areas compare?

Something to think about . . .

This rectangle is 7 squares long and 4 squares tall. The total area is 28 square units. The area of the parallelogram is also 28 square units.

BRAIN TICKLERS
Set # 45

Use this diagram to solve problems 1 and 2.

1. What is the perimeter of this parallelogram?

2. What is the area of this parallelogram?

3. What is the area of a parallelogram with base 6 and height 4?

(Answers are on page 248.)

TRAPEZOIDS

A *trapezoid* is a quadrilateral with only two parallel sides.

Perimeter

To find the perimeter of a trapezoid, just add all four sides.

> Perimeter of a trapezoid = Base 1 + Base 2 + Side 1 + Side 2

EXAMPLE:
Find the perimeter of this trapezoid.

The perimeter of this trapezoid is $3 + 5 + 8 + 12 = 28$ units.

Area

The formula for the area of a trapezoid is $\frac{1}{2}$ (sum of bases) $\times a$.
To find the area of a trapezoid:

- find the length of both bases
- find the length of the altitude

You know all three of these measurements.

To find the area of a trapezoid, follow these three painless steps.

Step 1: Add base 1 and base 2.

Step 2: Take $\frac{1}{2}$ of the result of Step 1.

Step 3: Multiply the result of Step 2 by the altitude.

$$\text{Area of a trapezoid} = \frac{1}{2}(b_1 + b_2)a$$

EXAMPLE:

Find the area of this trapezoid.

Step 1: First add base 1 to base 2.

$$7 + 9 = 16$$

Step 2: Take $\frac{1}{2}$ of the result of Step 1.

$$\frac{1}{2}(16) = 8$$

Step 3: Multiply the result of Step 2 by the altitude.

$$8(4) = 32$$

The area of the trapezoid is 32 square units.

BRAIN TICKLERS
Set # 46

Use the diagrams to solve the problems.

1. What is the perimeter of trapezoid *ABCD*?
2. What is the area of trapezoid *ABCD*?

3. What is the perimeter of trapezoid *WXYZ*?
4. What is the area of trapezoid *WXYZ*?

(Answers are on page 248.)

RHOMBUSES

A *rhombus* is a parallelogram with four equal sides. A square is a special kind of rhombus since a square is a parallelogram with four equal sides. A square also has four right angles.

Perimeter

To find the perimeter of a rhombus, add all the sides.

Perimeter of rhombus = Side 1 + Side 2 + Side 3 + Side 4

EXAMPLE:

Find the perimeter of a rhombus if each of the sides of the rhombus is 10 inches.

The perimeter is 10 + 10 + 10 + 10 or 40 inches.

```
        10
     ┌──────┐
10  /        / 10
   /        /
  └────────┘
      10
```

Area

To find the area of a rhombus, follow these two simple steps.

Step 1: Multiply the length of the diagonals together.
$\overline{AC}$ is a diagonal of the rhombus.
$\overline{BD}$ is a diagonal of the rhombus.

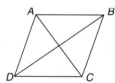

Step 2: Take one half of the answer.

$$\text{Area of a rhombus} = \frac{1}{2}(\text{Diagonal 1} \times \text{Diagonal 2})$$

EXAMPLE:

If one diagonal of a rhombus is 6 units long and the other diagonal is 5 units long, what is the area of the rhombus?

$$\text{Area} = \frac{1}{2}(6 \times 5) = \frac{1}{2}(30) = 15 \text{ square units}$$

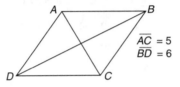

$\overline{AC} = 5$
$\overline{BD} = 6$

Experiment

Transform a rhombus into another common shape.

Materials
    Scissors
    Paper
    Pencil
    Ruler

Procedure

1. Draw a rhombus. Draw the diagonals of the rhombus.
2. Cut the rhombus along its diagonals. Four small triangles are formed.
3. Rearrange the four small triangles into a new shape. What shape can you make?

Something to think about . . .
    Is the area of a rhombus related to the area of any other quadrilateral?

REGULAR POLYGONS

A *regular polygon* is a polygon with equal sides and equal angles. An equilateral triangle and a square are both regular polygons. There are also regular pentagons, hexagons, heptagons, and octagons.

It is possible to construct a regular polygon with any number of sides.

Perimeter

To find the perimeter of a regular polygon, just add the sides. Or multiply the length of one side by the number of sides.

> Perimeter of a regular polygon = $n(s)$
> where n is the number of sides and
> s is the length of one side.

EXAMPLE:
Find the perimeter of a hexagon with side 2.

Add the sides of the hexagon together.

2 + 2 + 2 + 2 + 2 + 2 = 12
Or multiply the length of one side by the number of sides.
(2)6 = 12

Area

Finding the area of a regular polygon is easy.
Multiply $\frac{1}{2}$ the apothem times the perimeter.
The *apothem* is a line segment from the center
of the polygon perpendicular to a side.

Apothem

Area of a regular polygon = $\frac{1}{2}$ *AP*

where *A* is the apothem and *P* is the perimeter

To find the area of a regular polygon, follow these painless steps.

Step 1: Find the length of the apothem.

Step 2: Find the perimeter of the regular polygon.

Step 3: Multiply the apothem by the perimeter.

Step 4: Multiply the answer by $\frac{1}{2}$. The result is the area.

EXAMPLE:
Find the area of a hexagon with side 6 inches and apothem 5
inches.

Step 1: Find the length of the apothem. The apothem is 5 inches.

Step 2: Find the perimeter of the regular polygon. The perimeter
is 6(6) or 36.

Step 3: Multiply the apothem by the perimeter.

5(36) = 180

Step 4: Multiply the answer by $\frac{1}{2}$.

$\frac{1}{2}(180) = 90$ square inches

The area of the hexagon is 90 square inches.

Experiment

Compare two methods for finding the area of a square.

Materials
   Pencil
   Graph paper

Procedure

1. Draw a square with side 6.
2. Draw an apothem. Draw a line segment from the center of the square perpendicular to the opposite side.
3. Measure the length of the apothem.
4. Find the perimeter of the square.
5. Multiply $\frac{1}{2}$ of the apothem of the square by the perimeter to find the area of the square.
6. Now find the area of the same square by multiplying s times s.
7. Compare the areas of the square that you found by the two different methods.

Something to think about . . .
   What is the formula to find the perimeter and area of a regular octagon? An octagon is an eight-sided figure.

BRAIN TICKLERS
Set # 48

1. Find the perimeter of an octagon with sides 2 inches long.

2. Find the perimeter of a pentagon with sides 3 feet long.

3. Find the area of an octagon with sides 4 meters and an apothem 3 meters.

4. Find the area of a decagon with sides 3 inches long and apothem 5 inches. A decagon is a 10-sided figure.

(Answers are on page 248.)

SUPER BRAIN TICKLERS # 1

Look at each pair of figures. Determine which perimeter is larger or if they are both equal.

1. A rhombus with sides 4 feet long.
   A pentagon with sides 4 feet long.

2. A parallelogram with length 5 inches and width 10 inches.
   An equilateral triangle with sides 12 inches long.

3. A trapezoid with bases 4 and 8 centimeters and legs 3 and 9 centimeters.
   A square with sides 6 centimeters long.

Look at each pair of figures. Determine which area is greater or if they are both the same.

4. A rhombus with diagonals 4 feet long and 8 feet long.
   A triangle with base 4 feet long and height 8 feet long.

5. A rectangle with sides 3 and 5 miles long.
   A square with side 4 miles long.

6. A parallelogram with base 8 kilometers and height 4 kilometers long.
   A square with side 6 kilometers long.

(Answers are on page 248.)

Study Strategies

Write the names of each of the following shapes on a separate index card. Write the formula used to compute the area of each shape on the back of each card. Learn each of the following formulas using the flash cards.

Shape	Perimeter	Area
Triangle	Sum of all sides	$\frac{1}{2}b(h)$
Square	Sum of all sides	$s(s)$
Rectangle	Sum of all sides	$l(w)$
Parallelogram	Sum of all sides	$h(l)$
Trapezoid	Sum of all sides	$\frac{1}{2}(b_1 + b_2)h$
Rhombus	Sum of all sides	$\frac{1}{2}(d_1)(d_2)$
Regular polygon	Sum of all sides	$\frac{1}{2}(\text{apothem})(\text{perimeter})$
General	Sum of all sides	

Caution—Major Mistake Territory!

The area of any shape is always in square units. If you write the answer in units, it will be incorrect. The area of a square with a side 3 inches is 9 square inches, *not* 9 inches.

UNUSUAL SHAPES

To find the area of an unusual shape, add line segments to divide the shape into smaller known shapes. Find the area of each of these smaller shapes and add the result.

EXAMPLE:

Find the area of this shape. All the angles are right angles.

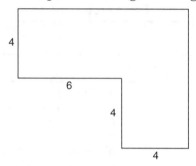

Add a line segment to change the shape into two rectangles.

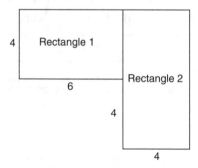

Find the area of rectangle 1 and rectangle 2.

> Rectangle 1 has a length of 6 and a width of 4. The area of rectangle 1 is 24 square units.
>
> Rectangle 2 has a length of 4 and a width of 8. The area of rectangle 2 is 32 square units.

Add the area of rectangle 1 and rectangle 2 together to find the areas of the entire shape. The area of the entire shape is 24 + 32, or 56 square units.

EXAMPLE:

Find the area of this shape.

Draw a line segment to divide this shape into a triangle and a square.

Find the area of the triangle.

$$\text{Area of triangle is } \frac{1}{2}(\text{Base}) \times (\text{Height})$$

The length of the height is 2 units and the length of the base is 4 units. The area of the triangle is $\frac{1}{2}(2)(4) = 4$ square units.

Find the area of the square.

$$\text{Area of square} = \text{Side} \times \text{Side}$$

The length of a side of the square is 4 units. The area is 16 square units.

Add the area of the square and the area of the triangle together to find the area of the shape.

$$4 + 16 = 20 \text{ square units}$$

The area of the shape is 20 square units.

BRAIN TICKLERS
Set # 49

Find the perimeter and area of this shape.

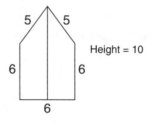

5   5

Height = 10

6   6

6

(Answers are on page 249.)

VOLUME

The volume of a solid figure is the capacity of the figure. The volume is the number of cubic units it contains. Cubes, boxes, cones, balls, and cylinders are all three-dimensional shapes. You can measure the volume of any of these shapes. The volume of three-dimensional shapes is measured in cubic units, such as cubic inches, cubic feet, cubic yards, cubic miles, cubic centimeters, cubic meters, or cubic kilometers. Cubic units are written by placing a small 3 over the units.

Experiment

Construct a cubic inch.

Materials
  Paper
  Pencil
  Ruler
  Scissors
  Scotch tape

Procedure

1. Copy the following diagram on a piece of paper. It is made out of six identical squares. Make each square 1 inch by 1 inch.
2. How many right angles are in the diagram?

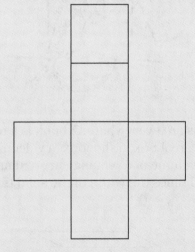

3. Cut out the diagram along the outside border.
4. Fold the diagram along the other edges and make a cube.

5. Tape the cube into place. This is one cubic inch.
6. Count the number of right angles on the cube.

Something to think about . . .
How many inches are in a foot?
How many cubic inches are in a cubic foot?

RECTANGULAR SOLIDS

Rectangular solids are everywhere. A book is a rectangular solid; so is a drawer, a cereal box, a shoebox, a videotape, and a CD case. To find the volume of a rectangular solid multiply the length of the solid times the width of the solid times the height of the solid.

Volume of a rectangular solid = $l \times w \times h$

EXAMPLE:

Find the volume of the rectangular solid with length 6, height 4, and width 4.

$4 \times 6 \times 4 = 96$ cubic units

CUBES

A *cube* is a special type of a rectangular solid. The length, width, and height of a cube are exactly the same.

Volume of a cube $= s \times s \times s = s^3$

EXAMPLE:

Find the volume of a cube with side 3 inches.

$3 \times 3 \times 3 = 27$ cubic inches
Notice that the answer is in cubic inches.

Surface Area

The surface area of a cube is six times the surface area of one side of the cube. Count the sides of a cube. There are six of them. The surface area of the cube is $6(s)(s)$.

Surface area of a cube $= 6s^2$

EXAMPLE:

What is the surface area of a cube with side 4?

$6(4)(4) = 96$ square units

Surface Area of a Rectangular Solid

The surface is the area on the outside of a three-dimensional shape. Imagine if you had to cover the entire outside of a three-dimensional shape with a piece of paper, how large would the piece of paper be? How could you compute the surface area of a three-dimensional shape?

Experiment

Find the surface area of a three-dimensional shape.

Materials
- Empty cereal box
- Ruler
- Scissors
- Paper
- Pencil

Procedure

1. Cut an empty cereal box along the edges of the box. You should cut the cereal box into six pieces. Each side of the box should be a separate piece.
2. Each of the pieces will be a rectangle. Measure the length and width of each rectangle. Enter the results in the chart.
3. Find the area of each of these rectangles by multiplying the length of each rectangle by the width of each rectangle. Enter the results in the chart.

Pieces of box	Length	Width	Area
Front of box			
Back of box			
Top of box			
Bottom of box			
Left side of box			
Right side of box			
Total surface area			

4. Add the areas of all six sides to find the total surface area of a cereal box.

Something to think about . . .
   Do any of the sides have the same area?
   How would you find the surface area of a pyramid?

To find the surface area of a rectangular solid, just follow these painless steps.

Step 1: Find the length, width, and height of the solid.

Step 2: Multiply 2 × length × width.

Step 3: Multiply 2 × width × height.

Step 4: Multiply 2 × length × height.

Step 5: Add the results of Steps 3, 4, and 5. The answer is the surface area of the rectangular solid.

EXAMPLE:
Find the surface area of a rectangular solid with sides 5 inches, 6 inches, and 7 inches.

Step 1: Find the length, width, and height of the solid.
The length is 5 inches, the width 6 inches, and the height 7 inches.

Step 2: Multiply 2 × length × width.

$2(5)(6) = 60$ square inches

Step 3: Multiply 2 × width × height.

$2(6)(7) = 84$ square inches

Step 4: Multiply 2 × length × height.

$2(5)(7) = 70$ square inches

Step 5: Add the results of Steps 3, 4, and 5. The answer is the surface area of the rectangular solid.

The surface area is 214 square inches.

Surface area of a rectangular solid =
$$2(\ell)(w) + 2(h)(w) + 2(\ell)(h)$$

Caution—Major Mistake Territory!

Surface area is measured in square units, not units. The surface area of a cube with side 3 inches is 54 square inches, *not* 54 inches.

BRAIN TICKLERS
Set # 50

1. What is the volume of a cube with side 5 inches?

2. What is the volume of a cube with side 1 inch?

3. What is the surface area of a cube with side 5 inches?

4. What is the surface area of a cube with side 1 inch?

5. What is the volume of a rectangular solid with sides 1, 2, and 3 inches?

6. What is the surface area of a rectangular solid with sides 1, 2, and 3 inches?

(Answers are on page 249.)

CYLINDERS

A *cylinder* is a common shape. A box of oatmeal is a cylinder, so is a glass, a can of green beans, or a can of tuna.

To find the surface area of a cylinder, follow these painless steps.

Step 1: Find the height of the cylinder.

Step 2: Find the radius of the cylinder.

Step 3: Find the area of the circle that is the base of the cylinder using the equation $A = (\text{pi})\, r^2$.

Step 4: Multiply the area found in Step 3 by 2 since there is a circle at both the top and the bottom of the cylinder.

Step 5: Find the circumference of the circle that forms the base of the cylinder. Use the formula $2(\text{pi})r$.

Step 6: Multiply the circumference of the circle (found in Step 5) by the height of the cylinder.

Step 7: Add the answer to Step 4 to the answer to Step 6.

Surface area of a cylinder = $2(\text{pi})\, r^2 + 2(\text{pi})\, rh$

EXAMPLE:

Find the surface area of a cylinder that is 12 inches high and has a diameter of 4 inches.

Step 1: Find the height of the cylinder.
The height is given as 12 inches.

Step 2: Find the radius of the cylinder.
Since the diameter of the cylinder is 4 inches, the radius of the cylinder is 2 inches.

Step 3: Find the area of the circle that is the base of the cylinder using the equation $A = (\text{pi})\, r^2$

$$A = \text{pi}(2)^2$$

$$A = 4\,\text{pi}$$

Step 4: Multiply the area found in Step 3 by 2 since there is a circle at both the top and the bottom of the cylinder.

Area of both circles = 2(4 pi) = 8 pi

Step 5: Find the Circumference of the circle that forms the base of the cylinder. Use the formula 2(pi)r.

Circumference = 2(pi)(2) = 4 pi

Step 6: Multiply the circumference of the circle (found in Step 5) by the height of the cylinder.

Area of sides of cylinder = (4 pi)(12) = 48 pi

Step 7: Add the answer to Step 4 to the answer to Step 6.

8 pi + 48 pi = 56 pi

The surface area of the cylinder is 56 pi.

Caution—Major Mistake Territory!

Think of the surface area of a cylinder as the area of the circle on the top of the cylinder plus the area of the circle on the bottom of the cylinder plus the area of the sides of the cylinder. All three areas must be included to get the correct area.

$$\text{Volume of a cylinder} = (\text{pi})\,r^2 h$$

To find the volume of a cylinder, follow these three painless steps:

Step 1: Square the radius.

Step 2: Multiply the result by pi, which is 3.14.

Step 3: Multiply the answer by the height.

EXAMPLE:

Find the volume of a cylinder with radius 5 and height 10.

Step 1: Square the radius.

$$5 \times 5 = 25$$

Step 2: Multiply 25 by pi, which is 3.14.

$$25 \times 3.14 = 78.5$$

Step 3: Multiply 78.5 by 10.

$$78.5 \times 10 = 785$$

The volume of the cylinder is 785 cubic units.

Experiment

Compare the volume of three glasses.

Materials
  3 glasses
  Water
  Pencil
  Paper
  Ruler
  Calculator

Procedure

1. Find three glasses of three different sizes.
2. Using the ruler, measure the radius of each glass.
3. Using the ruler, measure the height of each glass.
4. Compute the volume of each glass using the formula $(pi)r^2h$

Circle	Radius	Height	Volume
Circle 1			
Circle 2			
Circle 3			

5. Based on the volume you computed, rank order the glasses from smallest to largest.
6. Fill what you computed to be the smallest glass with water.
7. Pour the smallest glass of water into the next largest glass. Did all the water fit?
8. Fill this middle-size glass with water and pour it into the largest glass. Did all the water fit? Were all your calculations correct?

Something to think about . . .
  How would you compare the volume of a box of cereal to a glass of water?

BRAIN TICKLERS
Set # 51

1. Find the volume of a cylinder with radius 4 and height 2?

2. Find the volume of a cylinder with radius 1 and height 10?

3. Find the volume of a cylinder with radius 10 and height 1?

(Answers are on page 249.)

CONES

The volume of a cone is $\frac{1}{3}$ (pi)r^2h.

To find the volume of a cone, follow these painless steps:

Step 1: Square the radius.

Step 2: Multiply it by pi, which is 3.14.

Step 3: Multiply the answer in Step 2 by the height.

Step 4: Multiply the answer in Step 3 by $\frac{1}{3}$.

Notice that the formula for the volume of a cone is exactly one-third the volume of a cylinder of the same height.

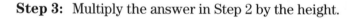

Volume of a cone = $\frac{1}{3}$ (pi) $r^2 h$.

EXAMPLE:

Find the volume of a cone with height 5 and diameter 6.

Step 1: Square the radius.

> The diameter of the cone is 6.
> The radius of the cone is half the diameter, or 3.
> 3^2 is 9.

Step 2: Multiply 9 by pi, which is 3.14.

$$9 \times 3.14 = 28.26$$

Step 3: Multiply 28.26 by the height of the cone, which is 5.

$$28.26 \times 5 = 141.3$$

Step 4: Multiply $141.3 \times \dfrac{1}{3} = 47.1$.

> The volume of the cone is 47.1 cubic units.

BRAIN TICKLERS
Set # 52

1. Find the volume of a cone with radius 6 and height 3.

2. Which has a greater volume, a cone with radius 2 and height 6 or a cylinder with radius 2 and height 2?

(Answers are on page 249.)

SPHERES

A ball, an orange, and a globe are all spheres. A *sphere* is the set of all points equidistant from a given point.

To find the surface area of a sphere, just follow these painless steps.

Step 1: Find the radius of the sphere.

Step 2: Square the radius.

Step 3: Multiply the answer by 4.

Step 4: Multiply the answer by pi. The answer is in square units.

$$\text{Surface area of a sphere} = 4(\text{pi})\,r^2$$

EXAMPLE:
Find the surface area of sphere with radius 10 inches.

Step 1: Find the radius of the sphere.
The radius is given as 10.

Step 2: Square the radius.

$$10^2 = 100$$

Step 3: Multiply the answer by 4.

$$4(100) = 400$$

Step 4: Multiply the answer by pi. The answer is in square units. The surface area of the sphere is 400 pi in.²

All you need to know to find the volume of a sphere is the radius.

To find the volume of a sphere, follow these three painless steps.

Step 1: Cube the radius. (Multiply $r \times r \times r$.)

Step 2: Multiply the answer by $\frac{4}{3}$.

Step 3: Multiply the answer by pi or by 3.14.

$$\text{Volume of a sphere} = \frac{4}{3}(\text{pi})r^3$$

EXAMPLE:

Find the volume of a sphere with radius 6.

Step 1: Cube the radius.

$$6 \times 6 \times 6 = 216$$

Step 2: Multiply the answer by $\frac{4}{3}$.

$$\frac{4}{3} \times 216 = 288$$

Step 3: Multiply the answer by pi.

The volume of the sphere is 288 pi units³.
Or, instead of multiplying by π, multiply the answer by 3.14.
288 × 3.14 = 904.32 cubic units

BRAIN TICKLERS
Set # 53

1. Compute the surface area of a sphere with radius 3.

2. Compute the volume of a sphere with radius 3.

3. Compute the surface area of a sphere with radius 1.

4. Compute the volume of a sphere with radius 1.

(Answers are on page 249.)

SUPER BRAIN TICKLERS # 2

Find the volume of these shapes. Use these formulas.

$$\text{Cube} = s^3$$

$$\text{Rectangular solid} = l \times w \times h$$

$$\text{Cylinder} = (\text{pi})r^2 h$$

$$\text{Cone} = \frac{1}{3}(\text{pi})r^2 h$$

$$\text{Sphere} = \frac{4}{3}(\text{pi})r^3$$

1. A sphere with radius 6.

2. A cone with height 4 and radius 2.

3. A sphere with diameter 10.

4. A cylinder with radius 4 and height 10.

5. A cube with side 8.

6. A rectangular solid with length 1, width 2, and height 3.

(Answers are on page 250.)

BRAIN TICKLERS—THE ANSWERS

Set # 42, page 206

1. 15 inches

2. 9 inches

3. 2 square inches

Set # 43, page 209

1. 32 inches

2. 40 meters

3. 50 square inches

4. 8 square centimeters

Set # 44, page 213

1. 28 units

2. 25 square units

3. 9 square units

4. 40 units

Set # 45, page 216

1. 44 inches
2. 72 square inches
3. 24 square units

Set # 46, page 219

1. 22 units
2. 18 square units
3. 28 units
4. 36 square units

Set # 47, page 222

1. 16 inches
2. 24 square inches
3. 1 square foot

Set # 48, page 225

1. 16 inches
2. 15 feet
3. 48 square meters
4. 75 square inches

Super Brain Ticklers # 1, page 225

1. A pentagon with sides 4 feet long is larger.
2. An equilateral triangle with side 12 inches is larger.
3. They are both the same.
4. They are both the same.
5. A square with side 4 miles is larger.
6. A square with side 6 kilometers is larger.

Set # 49, page 230

Perimeter = 28 units, Area = 48 square units

Set # 50, page 237

1. 125 cubic inches
2. 1 cubic inch
3. 150 square inches
4. 6 square inches
5. 6 cubic inches
6. 22 square inches

Set # 51, page 242

1. 32 pi cubic units
2. 10 pi cubic units
3. 100 pi cubic units

Set # 52, page 243

1. 36 pi cubic units
2. They are the same.

Set # 53, page 246

1. 36 pi square units or 113.04 square units
2. 36 pi cubic units
3. 4π square units
4. $\frac{4}{3}$ pi cubic units

Super Brain Ticklers # 2, page 247

1. 288 pi cubic units or 904.32 cubic units

2. $16\frac{1}{3}$ pi cubic units or 51.30 cubic units

3. $1333\frac{1}{3}$ pi cubic units or 4186.69 cubic units

4. 160 pi cubic units or 502.4 cubic units

5. 512 cubic units

6. 6 cubic units

Graphing

YOU ARE HERE

Points, lines, polygons, circles, and a variety of other figures can be graphed on a plane. To graph these figures, a grid is drawn. The grid, called a *coordinate plane*, shows where different points are located on the plane. Start by drawing a horizontal number line called the x-axis. Draw a vertical number line, called the y-axis, perpendicular to the x-axis.

Notice:

The x-axis and y-axis intersect at the origin.
Every point on the x-axis to the right of the y-axis is positive.
Every point on the x-axis to the left of the y-axis is negative.
Every point on the y-axis above the x-axis is positive.
Every point on the y-axis below the x-axis is negative.

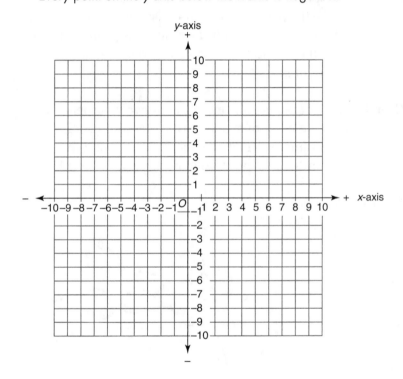

GRAPHING POINTS

Coordinate points are used to graph points on the coordinate plane. Coordinate points are written by placing two numbers, x and y, in parentheses, (x, y).

The first number, x, tells where the point is on the x-axis. Positive numbers are located to the right of the y-axis while negative numbers are located to the left of the y-axis. The second number, y, indicates where the point lies on the y-axis. Positive numbers move up above the x-axis, and negative numbers move down below the x-axis.

To graph a coordinate point, follow these four painless steps.

Step 1: Put your pencil at the origin.

Step 2: Look at the x-coordinate. If the number is negative, move the pencil to the left the same number of spaces as the x-coordinate. If the number is positive, move the pencil to the right the same number of spaces as the x-coordinate.

Step 3: Look at the y-coordinate, and move your pencil the same number of spaces as the y-coordinate. If the y-coordinate is negative move down, but if the y-coordinate is positive move up.

Step 4: Put a dot at this point.

EXAMPLE:
Graph the point (3, 1).

Step 1: Put your pencil at the origin.

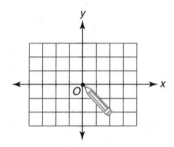

Step 2: Move your pencil right or left along the *x*-axis based on the *x*-coordinate.
Since the *x* coordinate is 3, move your pencil to the right 3 spaces.

Step 3: Move your pencil up or down based on the *y*-coordinate.
Since the *y*-coordinate is 1, move up 1 space.

Step 4: Put a dot at this point.
Label the point (3, 1).

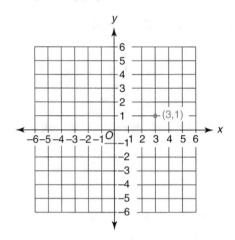

EXAMPLE:

Graph the point (0, –3).

Step 1: Put your pencil at the origin.

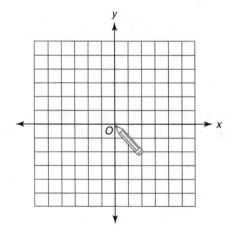

Step 2: Move your pencil right or left along the x-axis based on the x-coordinate. Since the x-coordinate is zero, keep your pencil at the origin.

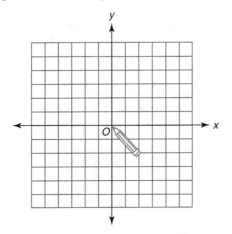

Step 3: Move your pencil up or down based on the y-coordinate. Since the y-coordinate is –3, move down three spaces.

Step 4: Put a dot at this point.
Label the point (0, –3).

BRAIN TICKLERS
Set # 54

Graph the following points on the coordinate axis.

1. (1, –3)
2. (–1, 2)
3. (4, 1)
4. (–2, –2)
5. (5, 0)

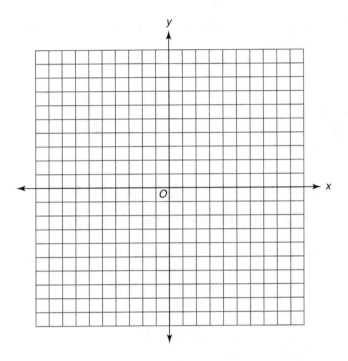

(Answers are on page 296.)

QUADRANTS

The coordinate plane is divided into four quadrants.

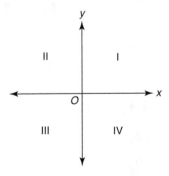

It's possible to determine in which quadrant a point is located just by looking at it.

Points that contain both a positive x and a positive y are located in quadrant I.

Points that contain a negative x and a positive y are located in quadrant II.

Points that contain both a negative x and a negative y are located in quadrant III.

Points that contain a positive x and a negative y are located in quadrant IV.

Caution—Major Mistake Territory!

The quadrants are counted counterclockwise, not clockwise. Just remember where the positive and negative numbers go and you will have no trouble with quadrants.

Quadrant I points have the form (+, +).
Quadrant II points have the form (–, +).
Quadrant III points have the form (–, –).
Quadrant IV points have the form (+, –).

BRAIN TICKLERS
Set # 55

In which quadrant are the following points located?

1. (–3, 4)

2. (2, 2)

3. (100, –4)

4. (–2, –2)

5. (0, 0)

(Answers are on page 296.)

THE MIDPOINT FORMULA

The midpoint of a line segment is the single point equidistant from both endpoints. It's easy to find the midpoint of the segment with the endpoints (0, 0) and (2, 0). Just graph the two points and connect them.

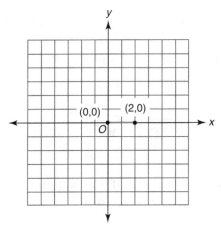

It's obvious that the midpoint is (1, 0) since this segment is just 2 units long. But finding the midpoint of other segments is not so easy. You have to use the midpoint formula.

Midpoint Formula

The midpoint of the line segment with endpoints (x_1, y_1) and (x_2, y_2) is

$$\frac{x_1 + x_2}{2}, \frac{y_1 + y_2}{2}$$

The midpoint formula uses subscripts, which are the small numbers written after the letters x and y. The small letters after the x and y are used to tell the difference between the different x's and y's used in the equation.

x_1 is read x sub one.
x_2 is read x sub two.
y_1 is read y sub one.
y_2 is read y sub two.

To find the midpoint of a line segment, just follow these painless steps.

Step 1: Find the endpoints of the line segment.

Step 2: Add the two x-coordinates and divide by 2.

Step 3: Add the two y-coordinates and divide by 2.

Step 4: The answers to Step 2 and Step 3 are the coordinates of the midpoint.

EXAMPLE:

Find the midpoint of the segment that connects the points $(3, 2)$ and $(1, 6)$.

Step 1: Find the coordinates of the endpoints of the segment. The endpoints are given as $(3, 2)$ and $(1, 6)$.

Step 2: Add the x-coordinates and divide by 2.
$$\frac{(3+1)}{2} = \frac{4}{2} = 2$$

Step 3: Add the y-coordinates and divide by 2.
$$\frac{(2+6)}{2} = \frac{8}{2} = 4$$

Step 4: The coordinates of the midpoint are $(2, 4)$.

EXAMPLE:

Find the midpoint of the segment that connects the points $(0, -3)$ and $(4, -1)$.

Step 1: Find the coordinates of the endpoints of the segment. The endpoints are given as $(0, -3)$ and $(4, -1)$.

Step 2: Add the x-coordinates and divide by 2.

$$\frac{(0+4)}{2} = \frac{4}{2} = 2$$

Step 3: Add the y-coordinates and divide by 2.

$$\frac{(-3+-1)}{2} = \frac{-4}{2} = -2$$

Step 4: The coordinates of the midpoint are (2, –2).

Experiment

Find the midpoint of the line segment.

Materials
    Graph paper
    Pencil
    Ruler

Procedure

1. Graph each of the line segments.
2. Estimate the midpoint of each segment by examining the line segment. Enter your estimate in the chart.
3. Determine the midpoint of each segment using the midpoint formula:

$$\frac{x_1 + x_2}{2}, \frac{y_1 + y_2}{2}$$

Enter the results in the chart.

Line segment	Endpoint 1	Endpoint 2	Estimate midpoint	Midpoint formula
Line segment 1	(3, 1)	(−1, 1)		
Line segment 2	(2, 8)	(2, 3)		
Line segment 3	(3, 5)	(1, 3)		
Line segment 4	(3, 3)	(−3, −3)		
Line segment 5	(−1, −4)	(2, 6)		

Something to think about . . .
   What is the difference between your estimate and using the midpoint formula?

BRAIN TICKLERS
Set # 56

Find the midpoint of each of the segments formed by connecting the following pairs of points.

1. (6, 4) and (−2, 6)

2. (2, 2) and (8, 8)

3. (−3, −3) and (5, 5)

4. (0, −2) and (4, 0)

(Answers are on page 297.)

THE DISTANCE FORMULA

The length of a line segment is the distance from one end of the segment to the other. Distance is measured in units (i.e., inches, feet, meters).

To find the length of a line segment, just follow these painless steps.

Step 1: Find the endpoints of the line segment.

Step 2: Subtract the smaller x-coordinate from the larger x-coordinate.

Step 3: Square the difference.

Step 4: Subtract the smaller y-coordinate from the larger y-coordinate.

Step 5: Square the difference.

Step 6: Add Step 3 and Step 5.

Step 7: Take the square root of Step 6. That's the distance from one end of the segment to the other.

> **Distance Formula**
> The distance between any two points $(x_1, y_1) + (x_2, y_2)$ is
> $$\sqrt{(x_2 - x_1)^2 + (y_2 - y_1)^2}$$

EXAMPLE:

What is the distance between the points $(5, 5)$ and $(1, 2)$?

Step 1: Find the endpoints of the line segment. The endpoints of the segment are given as $(5, 5)$ and $(1, 2)$.

Step 2: Subtract the smaller x-coordinate from the larger x-coordinate.

$$5 - 1 = 4$$

Step 3: Square the difference.

$$4 \times 4 = 16$$

Step 4: Subtract the smaller y-coordinate from the larger y-coordinate.

$$5 - 2 = 3$$

Step 5: Square the difference.

$$3 \times 3 = 9$$

Step 6: Add Step 3 and Step 5.

$$16 + 9 = 25$$

Step 7: Take the square root of Step 6.

$$\sqrt{25} = 5$$

The distance between the points $(1, 2)$ and $(5, 5)$ is 5.

Experiment

Find the length of a line segment using the Pythagorean theorem.

Materials
 Graph paper
 Pencil
 Ruler

Procedure

1. Draw a set of coordinate axes on a piece of graph paper.
2. Pick any two points and connect them.
3. Make a right triangle using the two points as two of the vertices of the triangle.
 Draw a line parallel to the x-axis through the point with the lower y value.
 Draw a line parallel to the y-axis through the point with the higher y value. Where the two lines intersect is the right angle of the right triangle.
4. Count the squares to determine the length of each side of the triangle. Enter the results in the chart.
5. Compute the length of the hypotenuse. Use the Pythagorean theorem. Enter the results in the chart.
6. Compute the distance between the two points using the distance formula. Enter the results in the chart.
7. Repeat steps 1 to 6, picking another set of 2 points.
8. Repeat steps 1 to 6 until the chart is complete.

Set	Point 1	Point 2	Length of horizontal side	Length of vertical side	Length of hypotenuse	Distance formula
Set 1						
Set 2						
Set 3						
Set 4						

Something to think about . . .
 Compare the answers you found using the distance formula to the length of the hypotenuse. Are they different?

Caution—Major Mistake Territory!

The distance between two points is always a positive number.

BRAIN TICKLERS
Set # 57

Find the distance between these two points.

1. $(4, 2)$ and $(1, 6)$

2. $(0, 6)$ and $(1, 4)$

3. $(-3, 2)$ and $(4, -1)$

4. $(-2, 0)$ and $(0, 1)$

(Answers are on page 297.)

GRAPHING A LINE BY PLOTTING POINTS

An equation of a line has x- and/or y-variables. The following equations are lines.

$$x + y = 9$$

$$3x + 2y = 1$$

$$y = -3x + 2$$

$$y = 4x$$

$$y = -5$$

$$x = 7$$

These are *not* equations of a line, since these equations have an x^2 term and/or a y^2 term.

$$3x^2 + 5y = 7$$

$$2x + 9y^2 = 1$$

$$3y^2 - 2y^2 = 4$$

There are several ways to graph a line. The easiest way is plotting points.

To plot points, just follow these five painless steps.

Step 1: Substitute 0 for x and solve for y.

Step 2: Substitute 1 for x and solve for y.

Step 3: Substitute –1 for x and solve for y.

Step 4: Graph the three points.

Step 5: Connect the three points to form a line. Label the line.

If the three points do not lie in a straight line, you have solved one of the equations incorrectly or graphed one of the three points incorrectly.

EXAMPLE:
Graph the equation $y = 3x - 2$.

Step 1: Substitute 0 for x and solve for y.

$$y = 3(0) - 2$$

$$y = -2$$

The first point is $(0, -2)$.

Step 2: Substitute 1 for x in the equation and solve for y.

$$y = 3(1) - 2$$

$$y = 1$$

The second point is (1, 1).

Step 3: Substitute –1 for x in the equation and solve for y.

$$y = 3(-1) - 2$$

$$y = -5$$

The third point is (–1, –5).

Step 4: Graph the points.

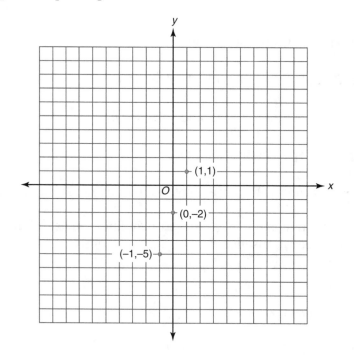

Step 5: Connect the points to form the line $y = 3x - 2$. Label the line.

BRAIN TICKLERS
Set # 58

Graph the following lines.

1. $y = 4x - 6$

2. $2x + 2y = 10$

3. $x = y$

(Answers are on page 297.)

GRAPHING HORIZONTAL AND VERTICAL LINES

Horizontal and vertical lines are exceptions to the graphing rule. Horizontal lines do not have an x term. They are written in the form y = some number. $y = 3$, $y = 0$, and $y = -7$ are all horizontal lines.

EXAMPLE:

Graph the equation $y = 4$.

Just plot points.
If x is 0, y is 4.
If x is 1, y is 4
If x is −1, y is 4.
No matter what number x is, y is always 4.
The graph of $y = 4$ is a horizontal line that intersects the y-axis at 4.

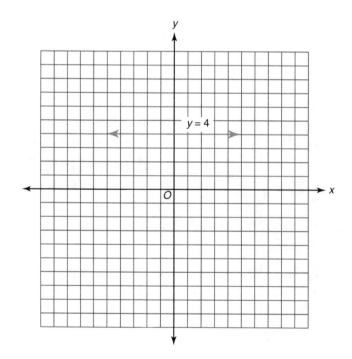

EXAMPLE:

Graph the line $y = -2$. Just draw a horizontal line at $y = -2$ parallel to the x-axis, since no matter what x is, $y = -2$.

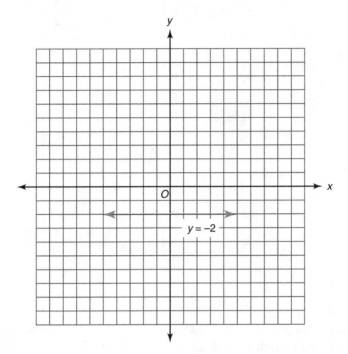

Vertical lines do not have a y term. They are parallel to the y-axis. These lines are parallel to the y-axis: $x = 1$, $x = -5$, and $x = 0$.

EXAMPLE:

Graph the equation $x = -2$.

Just plot points.
If $y = 1$, $x = -2$.
If $y = -1$, $x = -2$.
If $y = 0$, $x = -2$.

No matter what number y is, x is always -2.
The graph of $x = -2$ is a vertical line that intersects the x-axis at -2.

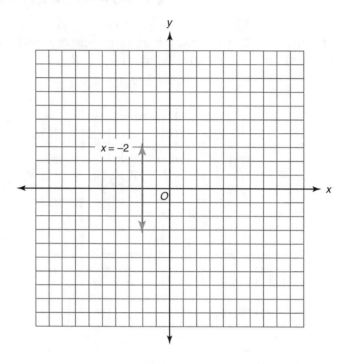

EXAMPLE:

Graph the equation $x = 5$. Just graph a line parallel to the y-axis at $x = 5$, since no matter what y is, x is always 5.

Caution—Major Mistake Territory!

Any line of the form $y = 3$ intersects the y-axis. It is not parallel to the y-axis.
Any line of the form $x = 3$ intersects the x-axis. It is not parallel to the x-axis.

BRAIN TICKLERS
Set # 59

Decide whether each of the following statements is true or false.

1. The line $y = 4$ is parallel to the x-axis.

2. The line $x = -1$ intersects the x-axis.

3. The line $y = 3$ intersects the y-axis.

4. The line $y = 0$ is the x-axis.

5. The line $x = 0$ is the x-axis.

6. The line $x = 3$ is parallel to the y-axis.

(Answers are on page 299.)

THE SLOPE

The *slope* of a line is a measure of the incline of the line. A positive slope indicates that the line goes uphill if you are moving from left to right. These lines have positive slopes.

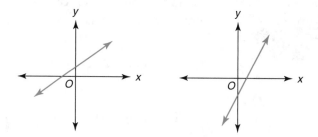

A negative slope indicates that a line goes downhill if you are moving from left to right. These lines have negative slopes.

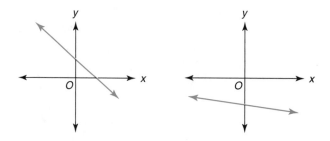

There are two common ways to find the slope of a line.

1. Putting an equation in slope-intercept form.

2. Using the point-point method.

Method 1: Slope-intercept method

To change an equation to slope-intercept form, just follow these two painless steps.

Step 1: Solve the equation for y.

Step 2: The variable in front of the x is the slope.

EXAMPLE:
Find the slope of the line $2x + y - 2 = 0$.

Step 1: Solve the equation $2x + y - 2 = 0$ for y.

Subtract 2x from both sides of the equation.

$$y - 2 = 2x$$

Add 2 to both sides of the equation.

$$y = 2x + 2$$

Step 2: The number in front of x is the slope.
2 is in front of x, so 2 is the slope.

EXAMPLE:
Find the slope of the equation $3x + 6y = 18$.

Step 1: Solve the equation $3x + 6y = 18$ for y.

Subtract 3x from both sides of the equation.

$$6y = -3x + 18$$

Divide both sides of the equation by 6.

$$y = -\frac{1}{2}x + 3$$

Step 2: The number in front of the x is the slope.

The slope is $-\frac{1}{2}$.

BRAIN TICKLERS
Set # 60

Change each of the following equations to slope-intercept form and find the slope.

1. $5x - y + 2 = 0$
2. $4x + 2y = -8$
3. $7y = -3x$
4. $x + 2y = 1$

(Answers are on page 299.)

Method 2: Point-point method

To find the slope of a line using the point-point method, you have to find two points on the line.

Using two points to find the slope of a line is painless.

Step 1: Find two points on the line.

Step 2: Subtract the first y-coordinate (y_1) from the second y-coordinate (y_2) to find the change in y.

Step 3: Subtract the first x-coordinate (x_1) from the second x-coordinate (x_2) to find the change in x.

Step 4: Divide the change in y (Step 2) by the change in x (Step 3). The answer is the slope of the line.

$$\text{Slope} = \frac{y_2 - y_1}{x_2 - x_1}$$

EXAMPLE:

Find the slope of a line through the points (3, 2) and (7, 1).

Step 1: Find two points on the line. Two points on the line are given: (3, 2) and (7, 1).

Step 2: Subtracting the second y-coordinate from the first y-coordinate, find the change in y.

$$2 - 1 = 1$$

Step 3: Subtracting the second x-coordinate from the first x-coordinate, find the change in x.

$$3 - 7 = -4$$

Step 4: Find the slope of the line by dividing the change in y (Step 2) by the change in x (Step 3).

$$\frac{1}{(-4)} = -\frac{1}{4}$$

The slope is $-\dfrac{1}{4}$.

Experiment

Estimate the slope of a line.

Materials
  Pencil
  Graph paper
  Ruler

Procedure

1. Graph the pairs of points shown in the chart.
2. Connect each pair of points to form a line.
3. Look at each line and try to estimate the slope.
   Enter your guess in the chart.

Use these tips for help.
- If the slope goes uphill, it is positive.
- If the slope goes downhill, it is negative.
- The steeper the slope, the higher the number.
- Flatter lines have slopes less than 1.

4. Compute the slope by finding the change in y divided by
   the change in x. Enter your results in the chart in the column
   labeled "Actual slope."
5. Compare your estimates to the actual slopes.

Line	Point 1	Point 2	Estimated slope	Actual slope
Line 1	$(1, 1)$	$(5, 5)$		
Line 2	$(0, 5)$	$(5, 0)$		
Line 3	$(0, 0)$	$(1, 5)$		
Line 4	$(-5, 1)$	$(5, -1)$		

Something to think about . . .

What would a line with slope $\frac{1}{10}$ look like compared to a line
with slope 10?

BRAIN TICKLERS
Set # 61

Find the slope of the line through the following points.

1. (3, 0) and (0, –2)
2. (7, 1) and (2, 6)
3. (0, 0) and (5, 3)
4. (–3, –2) and (1, 1)

(Answers are on page 299.)

GRAPHING USING SLOPE-INTERCEPT

The easiest way to graph a line is to use the slope-intercept method. The words *slope-intercept* refer to the form of the equation. When an equation is in slope-intercept form it is written in terms of a single y.

Equations in slope-intercept form have the form $y = mx + b$.

y is a variable.
m represents the number of x's the equation contains.
b stands for the number in the equation. It also tells you where the equation intercepts the y-axis.

Follow these six painless steps to graph an equation using the slope-intercept form.

Step 1: Put the equation in slope-intercept form. Express the equation of the line in terms of a single y. The equation should have the form $y = mx + b$.

Step 2: The number without any variable after it (the b term) is the y-intercept. The y-intercept is where the line crosses the y-axis. Make a mark on the y-axis at the y-intercept. If the equation has no b term, the y-intercept is 0.

Step 3: The number in front of the x is the slope. In order to graph the line, the slope must be written as a fraction. If the slope is a fraction, leave it as it is. If the slope is a whole number, place it over 1.

Step 4: Start at the y-intercept and move your pencil up the y-axis the number of spaces in the numerator of the slope.

Step 5: If the fraction is negative, move your pencil to the left the number of spaces in the denominator in the slope. If the fraction is positive, move your pencil to the right the number of spaces in the denominator of the slope. Mark this point.

Step 6: Construct a line that connects the point on the y-axis with the point where the pencil ended up.

> An equation in slope-intercept form has the form $y = mx + b$, where m is the slope and b is where the line intercepts the y-axis.

EXAMPLE:
Graph the line $2y - 4x = 8$.

Step 1: Put equation $2y - 4x = 8$ in slope-intercept form.

Express the equation in terms of a single y.
$2y - 4x = 8$

Add 4x to both sides.
$2y = 4x + 8$

Divide both sides of the equation by 2.
$y = 2x + 4$

Step 2: The number without any variable after it is the
y-intercept. The y-intercept is where the line crosses
the y-axis. Make a mark at the y-intercept. The
y-intercept is 4.

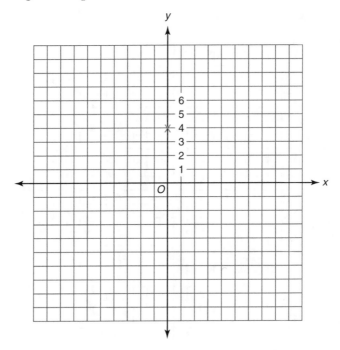

Step 3: The number in front of the x is the slope. If the slope is
a whole number, place it over 1. If it is a fraction, leave
it as it is. The slope is 2, so change it to $\frac{2}{1}$.

Step 4: Look at the numerator. Start at the y-intercept and move your pencil up the y-axis the number of spaces in the numerator. The numerator is 2, so move up two spaces to the point $y = 6$.

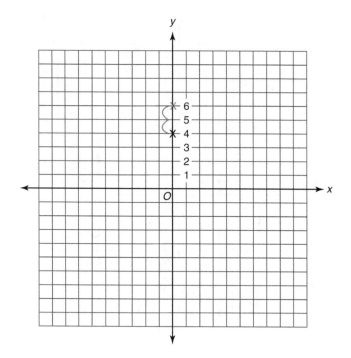

Step 5: Look at the denominator. If the fraction is negative, move your pencil to the left the number of spaces in the denominator. If the fraction is positive, move your pencil to the right the number of spaces in the denominator. The fraction is positive and the denominator is 1, so move your pencil one space to the right.

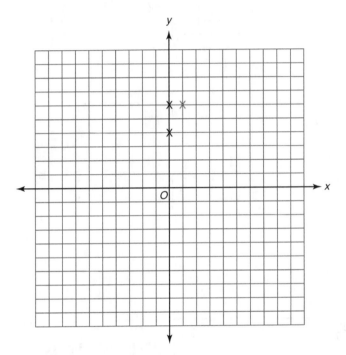

Step 6: Connect the point on the *y*-axis with the point where the pencil ended up.

BRAIN TICKLERS
Set # 62

What is the slope and *y*-intercept of each of these lines?

1. $y = 3x - 2$

2. $2y = 2x$

3. $4x - 4y = 8$

4. $2y - x = -10$

5. $y = 3$

(Answers are on page 299.)

FINDING THE EQUATION OF A LINE

If you know the slope of a line and a point on the line, you can find the equation of the line.

Just follow these four painless steps to find the equation of a line.

Step 1: Substitute the slope of the line for m in the equation $y = mx + b$.

Step 2: Substitute the coordinates of the point on the line for the variables x and y in the equation $y = mx + b$.

Step 3: Solve for b.

Step 4: Substitute m and b into the equation $y = mx + b$ to find the equation of a line.

EXAMPLE:
Find the equation of the line with slope -5 and through the point $(-2, -1)$.

Step 1: Substitute the slope, which is –5, for m in the equation $y = mx + b$.

$$y = -5x + b$$

Step 2: Substitute the coordinates of the point on the line for the variables x and y in the equation $y = mx + b$.

$$(-1) = -5(-2) + b$$

Step 3: Solve for b.

$$(-1) = (-5)(-2) + b$$
$$(-1) = 10 + b$$
$$b = -11$$

Step 4: Substitute m and b into the equation $y = mx + b$ to find the equation of a line. The equation of the line is

$$y = -5x - 11$$

BRAIN TICKLERS
Set # 63

Find the equation of each of these lines.

1. Slope = 2 and point (4, –4)
2. Slope = 1 and point (1, 1)
3. Slope = $\frac{1}{2}$ and point (0, –1)

(Answers are on page 300.)

TWO-POINT METHOD

If you know any two points, you can also find the equation of the line.

Just follow these four painless steps to find the equation of a line.

Step 1: Find the slope of the line. Divide the *change in y* by the *change in x*.

Step 2: Substitute the slope (m) into the equation $y = mx + b$.

Step 3: Substitute one pair of coordinates into the equation for x and y and solve for b.

Step 4: Substitute m and b into the equation $y = mx + b$ to find the equation of a line.

EXAMPLE:
Find the equation of the line through the points (3, 4) and (6, 2).

Step 1: Find the slope of the line.

Subtract one pair of coordinates from the other set of coordinates.
The change in y is 4 – 2 or 2.
The change in x is 3 – 6 or –3.
Divide the change in y by the change in x to find the slope.
The slope is $-\dfrac{2}{3}$.

Step 2: Substitute the slope (m) into the equation $y = mx + b$.

$$y = -\frac{2}{3}x + b$$

Step 3: Substitute one pair of coordinates into the equation for x and y and solve for b.

$$4 = -\frac{2}{3}(3) + b$$

$$4 = -2 + b$$

$$b = 6$$

Step 4: Substitute m and b into the equation $y = mx + b$. As a result, the equation of a line through the points $(3, 4)$ and $(6, 2)$ is

$$y = -\frac{2}{3}x + 6$$

BRAIN TICKLERS
Set # 64

Find the equation of the line through each pair of points.

1. $(1, 1)$ and $(6, 2)$

2. $(0, 5)$ and $(2, 0)$

3. $(3, -1)$ and $(4, -2)$

(Answers are on page 300.)

PARALLEL AND
PERPENDICULAR LINES

Experiment

Learn the relationship between lines with the same slope.

Materials
    Graph paper
    Pencil
    Ruler

Procedure

1. Graph all three of the following equations on the same pair of coordinate axes.

$$y = 2x + 3$$

$$y = 2x - 1$$

$$2y = 4x + 2$$

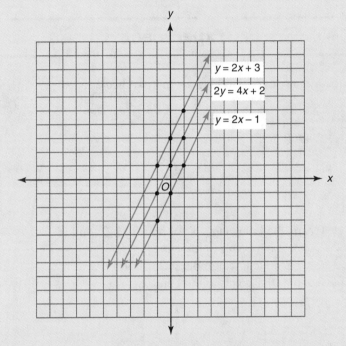

2. What do you notice about the lines you graphed?

Something to think about . . .

What do the equations you graphed have in common?
What do the equations $y = -x$, $y = -x + 1$, and $y = -x + 4$ have in common?

Theorem: If two lines are parallel, then they have the same slope.

Theorem: If two lines have the same slope, then they are parallel.

EXAMPLE:

$y = 2x + 1$ and $y = 2x - 6$ are parallel since they both have a slope of 2.

Experiment

Discover the relationship between perpendicular lines.

Materials
> Graph paper
> Pencil
> Ruler

Procedure

> 1. Graph the following pair of points (0, 0) and (3, 3).
> 2. Connect the points to form line 1.
> 3. Graph the following points (3, 0) and (0, 3).
> 4. Connect the points to form line 2.

> 5. Line 1 and line 2 should be perpendicular to each other.
> 6. Determine the slope of line 1.
> 7. Determine the slope of line 2.

8. Multiply the two slopes together. What's your answer?
9. On a second set of coordinate axes draw line 1 through the points (0, 1) and (1, 3).
10. Draw line 2 through the points (0, 1) and (2, 0).
11. Line 1 should be perpendicular to line 2.
12. Find the slope of both these lines and multiply them together. What's your answer?
13. Do you notice a pattern?

Something to think about . . .
How can you tell if two lines are perpendicular to each other?

Theorem: If two lines are perpendicular, then the slope of one line is the negative reciprocal of the slope of the other line.

Theorem: If the slope of one line times the slope of another line is −1, then the lines are perpendicular.

EXAMPLE:

$y = -3x + 2$ and $y = \frac{1}{3}x - 1$ are perpendicular lines.

Multiply the slope of the first line (−3) times the slope of the second line $\left(\frac{1}{3}\right)$. The answer is (−1).

BRAIN TICKLERS
Set # 65

Look at the following pairs of equations. Determine whether each pair of lines is perpendicular, parallel, or neither.

1. $y = 3x - 3$ and $y = -\dfrac{1}{3}x + 3$

2. $y = 2x + 5$ and $y = 2x + 7$

3. $y = -4x + 1$ and $y = -6x - 2$

4. $2y = x + 1$ and $y = \dfrac{1}{2}x + 3$

(Answers are on page 300.)

Study Strategies

Put the name of each of the following equations on one side of an index card and the formula on the other side. Memorize these formulas!

- Midpoint formula
- Distance formula
- Horizontal line
- Vertical line
- Slope
- Slope-intercept form
- Parallel lines
- Perpendicular lines

SUPER BRAIN TICKLERS

1. What quadrant is the point $(3, -2)$ located in?

2. What is the midpoint of the segment with endpoints $(3, 2)$ and $(-7, 0)$?

3. What is the distance between the points $(-1, -1)$ and $(4, 5)$?

4. Put the line $2x + 10y = 40$ in slope-intercept form.

5. Find the slope of the line through the points $(5, 1)$ and $(-1, -2)$.

6. Find the equation of the line with slope 4 and intercept $(0, 0)$.

7. At what point does the line $y = 4$ cross the y-axis?

8. At what point does the line $x = 4$ cross the y-axis?

9. What is the equation of the y-axis?

10. What is the equation of the line through the points $(3, 5)$ and $(1, 8)$?

11. What is the relationship between the lines $y = -\dfrac{1}{2}x + 1$ and $y = 2x$?

12. What is the relationship between the lines $10y = -5x + 4$ and $y = -\dfrac{1}{2}x - 1$?

(Answers are on page 300.)

BRAIN TICKLERS—THE ANSWERS

Set # 54, page 258

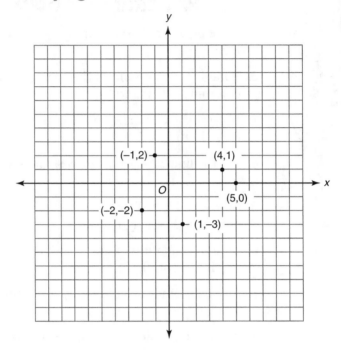

Set # 55, page 260

1. II
2. I
3. IV
4. III
5. The origin

Set # 56, page 264

1. $(2, 5)$

2. $(5, 5)$

3. $(1, 1)$

4. $(2, -1)$

Set # 57, page 268

1. 5 units

2. $\sqrt{5}$ units

3. $\sqrt{58}$ units

4. $\sqrt{5}$ units

Set # 58, page 271

1.

2.

3.

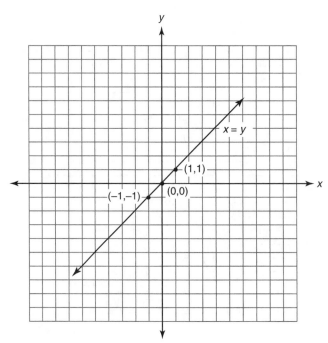

Set # 59, page 275

1. True

2. True

3. True

4. True

5. False

6. True

Set # 60, page 278

1. $y = 5x + 2$, slope $= 5$

2. $y = -2x - 4$, slope $= -2$

3. $y = -\dfrac{3}{7}x$, slope $= -\dfrac{3}{7}$

4. $y = -\dfrac{1}{2}x + \dfrac{1}{2}$, slope $= -\dfrac{1}{2}$

Set # 61, page 281

1. $\dfrac{2}{3}$

2. -1

3. $\dfrac{3}{5}$

4. $\dfrac{3}{4}$

Set # 62, page 286

1. Slope $= 3$, y-intercept $= -2$

2. Slope $= 1$, y-intercept $= 0$

3. Slope $= 1$, y-intercept $= -2$

4. Slope $= \dfrac{1}{2}$, y-intercept $= -5$

5. Slope $= 0$, y-intercept $= 3$

Set # 63, page 288

1. $y = 2x - 12$
2. $y = x$
3. $y = \dfrac{1}{2}x - 1$

Set # 64, page 290

1. $y = \dfrac{1}{5}x + \dfrac{4}{5}$
2. $y = -\dfrac{5}{2}x + 5$
3. $y = -x + 2$

Set # 65, page 294

1. Perpendicular
2. Parallel
3. Neither
4. Parallel

Super Brain Ticklers, page 295

1. Quadrant IV
2. $(-2, 1)$
3. $\sqrt{61}$
4. $y = -\dfrac{1}{5}x + 4$
5. $\dfrac{1}{2}$
6. $y = 4x$
7. $(0, 4)$
8. Never
9. $x = 0$
10. $y = -\dfrac{3}{2}x + 9\dfrac{1}{2}$
11. Perpendicular
12. Parallel

APPENDIX—
GEOMETRY RESOURCES
ON THE WEB

A History of Pi

http://www-groups.dcs.st-and.ac.uk/~history/HistTopics/
Pi_through_the_ages.html
A History of Pi tells you everything you ever wanted to know
about pi.

Ask Dr. Math

http://forum.swarthmore.edu/dr.math/
You can *Ask Dr. Math* a math question and get an answer or you
can scroll through thousands of math questions and see how
they are solved.

Boxer Math

http://www.boxermath.com
Boxer Math is a comprehensive learning environment for math.
Boxer Math offers high-technology tutorials in geometry and
algebra.

Color Math Pink

http://www.colormathpink.com
Color Math Pink is designed to help girls improve their math skills and reduce the math gender gap. The site provides help in all areas of mathematics from 4th to 12th grade, including geometry. At *Color Math Pink* you can take a diagnostic test in geometry or find a geometry tutor.

Dave's Math Tables

http://www.sisweb.com
Dave's Math Table is in English and Spanish and provides math lessons and help in a variety of areas, including geometry.

Math Archives

http://archives.math.utk.edu
Math Archives contains teaching materials for kindergarten through 12th grade, contests, and competitions.

Math Forum

http://forum.swarthmore.edu/
Run by Swarthmore College, the *Math Forum* is the most comprehensive math site on the web. It features a reference shelf to help with homework, chat areas, math games, contests, and over 7,500 math links.

Mega Mathematics

http://www.c3.lanl.gov/mega-math/
Developed by the Los Alamos National Laboratory, *Mega Mathematics* provides games and activities to improve the mathematics performance of elementary and middle school students.

Miss Glosser's Math Goodies

http://www.mathgoodies.com
Miss Glosser's Math Goodies contains interactive math activities on a variety of topics on everything from pre-algebra to probability. Also included is a math chat area, puzzles, and practice worksheets.

Ole Miss: Problems of the Week

http://www.olemiss.edu/mathed/problem.htm
The *Problem of the Week* is an educational web site that originates at the University of Mississippi. Problems are posted in four different subjects on a weekly basis. Prizes are offered by Casio to randomly selected winners.

S.O.S. Mathematics

http://www.sosmath.com
S.O.S. Mathematics contains more than 2,500 math pages filled with short and easy-to-understand explanations from simplifying fractions to differential equations. *S.O.S. Mathematics* also contains CyberExams to help you prepare for a test, and a CyberBoard where you can ask questions. *S.O.S. Mathematics* contains links to other sites on the web.

Think Quest

http://library.thinkquest.org/2647/geometry/geometry.htm
Think Quest provides explanations about various topics in geometry. This site has a special page on constructions and plenty of sample problems to help you understand the concepts. *Think Quest* also has a huge glossary.

INDEX

NOTES